Architectural Design in Steel

is the Steel Construction Institute (SCI). Its overall objective is to promote and develop the proper and effective use of steel. It achieves this aim through research, development of design aids and design approaches, publications and advisory and education services. Its work is initiated and guided through the involvement of its members on advisory groups and technical committees. The SCI is financed through subscriptions from its members, revenue from research and consultancy contracts and by sales of publications.

Membership is open to all organisations and individuals that are concerned with the use of steel in construction, and members include designers, architects, engineers, contractors, suppliers, fabricators, academics and government departments in the United Kingdom, elsewhere in Europe and in countries around the world. A comprehensive advisory and consultancy service is available to members on the use of steel in construction.

Further information on membership, publications and courses is given in the SCI prospectus available free on request from:

The Membership and Council Secretary
The Steel Construction Institute
Silwood Park
Ascot
Berkshire
SL5 7QN
Telephone: 01344 23345
Fax: 01344 22944
Website: www.steel-sci.org

Corus (formerly British Steel) sponsored the preparation of this book by the SCI and this support is gratefully acknowledged. The different divisions of Corus produce and market a comprehensive range of steel products for construction. Advisory services are available to help specifiers with any problems relevant to structural steelwork and to provide points of contract with the sales functions and technical services. A series of publications is available dealing with steel products and their use. A list of addresses and telephone numbers is given in Chapter 16.

Architectural Design in Steel

Peter Trebilcock and Mark Lawson

Routledge
Taylor & Francis Group
LONDON AND NEW YORK

The Steel Construction Institute

First published 2004 by Spon Press

Published 2016 by Routledge
2 Park Square, Milton Park, Abingdon, Oxon OX14 4RN
711 Third Avenue, New York, NY 10017, USA

First issued in hardback 2015

Routledge is an imprint of the Taylor and Francis Group, an informa business

Designed and typeset by Alex Lazarou, Surbiton, Surrey

British Library Cataloguing in Publication Data
A catalogue record for this book is available from the British Library

Library of Congress Cataloging in Publication Data
Trebilcock, Peter.
Architectural design in steel/Peter Trebilcock and Mark Lawson.
p. cm.
Includes bibliographical references.
ISBN-13: 978-0-419-24490-5 (pbk)

1. Building, Iron and steel. 2. Steel framing (Building) I. Lawson, R. M. II. Title.
TH1611.T697 2003
721'.04471—dc21

2003005976

ISBN-13: 978-1-138-13626-7 (hbk)

Contents

Preface

Architectural Design in Steel presents general design principles and examples of good practice in steel design, fabrication and architectural detailing. The book covers three areas:

- general principles of steel design
- opportunities for architectural expression
- examples of details used in recent projects.

The book includes all aspects of the architectural uses of steel in internal and external applications. The different types of structural members, frames and their connections are identified, and common details are discussed. Examples of the expressive use of steel are presented, including arches, tension structures, masts and glazing support systems. Connections between members, especially tubular connectors and cast steel nodes, are covered in detail.

Technical information is provided on fire and corrosion protection, and on penetrations through the building envelope. Reference is also made to other publications for more detailed guidance. Chapter 10 was drafted prior to the introduction of revised UK building regulations dealing with cold bridging. Designers are advised to seek specialist advice, where necessary, should steelwork penetration of the envelope be necessary.

The book was prepared by Peter Trebilcock, Consultant Architect to the Steel Construction Institute (SCI) and Head of Architecture at Amec Group Ltd, and by Mark Lawson, SCI Professor of Construction Systems at the University of Surrey (formerly Research Manager at the SCI). The work was funded by Corus (formerly British Steel (Sections, Plates and Commercial Steels)) and Corus Tubes and Pipes, and the former Department of the Environment, Transport and the Regions under the Partners in Technology initiative.

The assistance of the following individuals and organisations is acknowledged: Paul Salter, Consultant Structural Engineer; Christopher Nash, Nicholas Grimshaw and Partners; Rod McAllister, formerly Liverpool University School of Architecture; Paul Craddock, Arup; Eric Taylor, Arup; Craig Gibbons, Arup; Rob Watson, Foster and Partners; Geoff Hume, William Cook Steel Castings Limited; Alan Jones, Anthony Hunt Associates Ltd; John Pringle, Pringle Richards Sharratt; David Cash, Building Design Partnership; Michael Powell, Amec Group Ltd; Alan Ogg, The Royal Australian Institute of Architects. Information on tension cables was provided by Guy Linking Ltd.

Illustration credits

Angle Ring: 4.16
Arup: 7.37, 8.6, 13.7
Michael Barclay Partnership: 6.32
Barnshaws: 4.14
Benthem & Crouwel: 6.1
David Bower: 7.1
The British Architectural Library, RIBA: 3.4
Richard Bryant/ARCAID: 2.17, 9.16, 11.10
Canadian Institute of Steel Construction: 4.20
Martin Charles/VIEW: 2.6, 3.14
Classen: 7.9
Ian Clook: 2.5
Peter Cook/VIEW: 10.4
Corus: 4.25
Corus Tubes: 6.9, 6.10, 6.21, 6.40, 11.1
John Critchley: 3.12
Frank Dale Ltd: 3.11
Brian Davenport: 2.15
Richard Davies/Foster and Partners: 2.11, 7.15, 7.41
Richard Davis/VIEW: 7.5
M. Denance: 4.27
Jeremy Dixon.Edward Jones: 4.28
Peter Durant/Archblue.com: 1.6
Fabsec Ltd: 13.3
Norman Foster: 2.3
Foster and Partners: 2.2a, 2.2b, 2.16, 8.8
K. Frahm: 3.5
Berengo Gardin: 13.8b
Greg Germany: 3.19
Dennis Gilbert: 7.6
Dennis Gilbert/VIEW: 4.17, 4.40, 7.3, 7.36, 8.4
Nicholas Grimshaw and Partners: 2.13
W. D. Gericke: 1.12
Goodwin Steel: 11.9
Martine Hamilton Knight: 1.7, 9.4, 9.14,
Bill Hastings/Arc Photo: 2.4
Alastair Hunter: 1.2, 6.30
Keith Hunter: 1.5
S. Ishida: 13.8c, 13.8d
David Jewell: 4.38
A. Keller/artur: 9.10
Ken Kirkwood: 1.1, 4.29, 7.2, 7.4, 7.38
Serge Kreis/Camerzindgrafen Steiner: 2.1
Ian Lawson: 8.3, 10.7
W&J Leigh: 13.2
H. Leiska: 9.12
D. Leistner/artur: 7.6
Lindapter International: 5.9
J. Linden: 7.8
J. Linden/ARCAID: 2.8, 7.8
McCalls Special Projects: 9.8
Raf Madka/VIEW: 1.14
T. J. S. Marr: 12.1
Hugh Martin: 4.26
David Moore: 6.11
Brian Pickell: 4.20
Jo Reid and John Peck: 1.8,1.1,2.9, 2.12, 4.34, 7.7, 9.3, 10.5, 12.2
Norlaki Okabe: 13.8a
Price & Myers: 4.37, 6.16, 7.35
L. R. Shipsides: 1.9
Timothy Soar: 1.4, 4.19
Tim Street-Porter: 3.7
Kees Stuip: 4.22
Rupert Truman: 1.10, 2.10
Jocelyne van den Bossche: 4.21
Morley von Sternberg: 4.36
Usinor: 11.7
Westok: 4.15

Colour Section
David Bower: 25
Peter Cook/FaulknerBrowns: 20
Peter Cook/VIEW: 18, 24
Graham Gaunt/Arup: 5
W. D. Gericke: 9
Dennis Gilbert/VIEW: 7, 21
Richard Glover: 10
Andrew Holt/VIEW: 23
Nicholas Kane/BPR: 19
Serge Kreis/Camerzindgrafen Steiner: 27
Lenscape: 8
Duccio Malgamba: 26
Peter McInven/VIEW: 14,16
Morley von Sternberg/WilkinsonEyre: 22
National Maritime Museum: 4
Jo Reid and John Peck: 13
Katsuhisa Kida: 15
Timothy Soar: 12
Jocelyne van den Bossche: 6
Nigel Young/Foster and Partners: 1, 2a, 2b, 3
Hodder Associates: 17
Axel Weiss: 11

Cover illustration
Jocelyne van den Bossche

All photographs not specifically credited are courtesy of the authors and line drawings are courtesy of The Steel Construction Institute.

The authors and publishers would like to thank the above individuals and organizations for permission to reproduce material. We have made every effort to contact and acknowledge copyright holders, but if any errors or omissions have been made we would be happy to correct them at a later printing.

A number of illustrations have been adapted from the publication by Alan Ogg, *Architecture in Steel: The Australian Context*, The Royal Australian Institute of Architects, 1987.

Chapter 1

Introduction

Pier Luigi Nervi said:

> 'A technically perfect work can be aesthetically inexpressive but there does not exist, either in the past or the present, a work of architecture which is accepted and recognised as excellent from the aesthetic point of view which is not also excellent from the technical point of view. Good engineering seems to be a necessary though not sufficient condition for good architecture.'

Good detailing is a function of the spatial arrangement of the elements, their slenderness and lightness, and the connections between them. Figures 1.1–1.11 illustrate good examples of steel detailing in a variety of structural applications.

The need for guidance on detailing

Steelwork offers the opportunity for architectural expression, as well as being a structurally versatile and adaptable material. Good quality detailing is vital because it affects structural performance, cost, buildability and, perhaps most importantly, appearance.

Whilst the choice of the structural form is often the province of the structural engineer, architects should have a broad appreciation of the factors leading to the selection of the structure and its details. Traditionally, most detailing of connections is the responsibility of the steelwork fabricator but, for exposed steelwork, detailing is of much more interest to the architect, as it impacts on the aesthetics of the structure.

In this respect it is important that designers appreciate the common fabrication and erection techniques which may exert a strong influence on the method and approach to the detailing of modern steelwork in buildings.

Connections to other materials

The attachments of other elements, such as cladding and stairs to the steel structure, are described in another series of publications. These 'interfaces' are crucial to the efficiency and buildability of steel-framed buildings. Reference is made to good practice details in the Steel Construction Institute's (SCI's) publications on curtain walling,[1] connections to concrete,[2] and lift-shaft details.[3]

1.1 Advantages of steel construction

The distinct advantages of the use of steel in modern building construction may be summarised as follows:

- The modular nature of its fabrication (a 'kit of parts'), which can be delivered 'just in time' to site when required.
- The potential for rapid erection of the framework on site, which also reduces local disruption, noise and site storage.
- It is prefabricated to a high degree of accuracy.
- Long spans can be achieved economically by a variety of structural systems in steel and composite construction, permitting greater usable space.
- Steel or composite frames are lighter than concrete frames of the same span, thus reducing foundation costs.
- Steelwork permits adaptation in the future, and components can be re-used by unbolting.
- Composite steel-concrete floors can contribute to a thermally efficient building.
- A high proportion of steel production is recycled from scrap, and all steel is recyclable.

1.2 Opportunity for architectural expression

Steelwork possesses various advantages for architectural expression, as follows:

- External structures clearly express their function.
- Slender members can be designed efficiently, particularly using tubular sections.
- 'Lightness' can be accentuated by openings in beams and by latticework in the form of trusses.
- Curved members, such as arches, can be formed easily.
- Tension structures are efficient and lightweight, particularly for long-span enclosures.
- Connections can be designed expressively.
- The fire resistance of exposed steelwork can be enhanced by the use of intumescent coatings, or by concrete or water filling (of tubular sections).
- Colours and finishes of painted steelwork can be used to great effect.

In architecture, the decision to express or conceal the structural frame, either externally or internally, is usually decided by aesthetic preference coupled with technical and functional issues. The desire to express the structure of the building is an association extending from the use of iron and early steel in the last century.

Having decided to express the structure, the architect then considers a number of design factors against which he may test his proposals. Such considerations may include architecture and functional, planning or organisational requirements, as follows.

Architectural requirements (Colour Plates 8, 13, 15, 16, 22 and 24):

- The required overall visual effect of solidity or transparency; multiplicity of elements or minimalism; individuality or repetition of elements.
- Use of bespoke or standardised components.
- The nature of the architectural language; i.e. elegance and slenderness; strength and robustness.
- The relationship in visual and functional terms between the inside and outside spaces.

Functional requirements (Colour Plates 3, 4, 5, 9 and 27):

- Building form and function.
- Dimensional parameters, i.e. height of building, scale, use of column-free space.
- Stability requirements (particularly for tall buildings).
- Initial cost and life-cycle cost.
- Climate; both internally and externally.
- Services provision and maintenance, and opportunities for services integration.
- Interface details, particularly of the cladding to the frame.
- Durability, including maintenance implications and time to first maintenance.
- Fire-safety considerations.
- Health and safety requirements are now extended to Construction (Design and Maintenance) Regulations 1994 (CDM Regulations) requirements.[4]
- Protection from impact damage and vandalism.

Planning or organisational requirements (Colour Plates 2, 10, 11 and 23):

- Local planning and statutory requirements, including building height, and impact of the building on the locality.
- Programme/timescale requirements, not only of the construction project, but also of the resources/demands placed on consultants.
- Agreement on the responsibilities of the architect, structural engineer and constructor.
- Client input and acceptability of the design concept.
- Availability of suitable resources for construction, and opportunities for prefabrication (e.g. on a remote site).

Excellent examples from the 1980s showed what could be achieved in the expressive use of steel. In the Sainsbury Centre, a simple portal-frame structure was proposed initially, but rejected in favour of the deeper and more highly articulated structural frame that was finally adopted (see Figure 1.1).

The highly perforated structural members of the Renault Parts Distribution Centre (Figure 1.2) are an 'architectural' expression of engineering and technological efficiency, yet they do not necessarily represent the most efficient structural solution. These are conceptual

1.1 Portal-frame structure used in the Sainsbury Centre, Norwich (architect: Foster and Partners)

issues in which both the structural engineer and the architect should share a close interest, and which must be resolved jointly at the early stages of design. However, many examples of exposed steel follow a much more straightforward approach (see Colour Plate 26).

1.3 Holistic approach

1.2 The Renault Parts Distribution Centre, Swindon, showing mast and tension structure (architect: Foster and Partners)

To achieve economic and practical architectural details, there has to be a basic appreciation of the performance of the overall structure itself and the loading conditions imposed on the member or component in question. The form of the structure will strongly influence the details employed.

For the architect, details often evolve through the logical stages of conceptual design, followed by further rationalisation into the detailed design, i.e. from the macro to the micro. The architect may approach the concept design with the key component details already in mind. However, the final solution will be influenced by structural issues, an understanding of the fabrication and construction process, and other functional constraints (see Colour Plate 14).

As one of the first examples of external support using masts and cables, the steelwork for the Renault Parts Distribution Centre (see Figure 1.2) was subject to considerable refinement at the design stage by computer analysis, and the components were finally rigorously tested to assess their load capacity. The 'mast and arm' details reached a high level of sophistication, creating a strong aesthetic and functional appeal for what could have been a bland enclosure.

1.4 Scale and ornament

1.4.1 Scale

Buildings should be designed well at a range of scales. An understanding and an appreciation of all the scales will help in the art of assembly and detailing. Therefore, an elegant and well-proportioned building will have been successfully considered at the large scale as well as in its details. Good details alone do not necessarily lead to architectural success. This achievement relies on the consideration of all elements of the composition of varying scale (see Colour Plates 14 and 20).

For example, the canopy of the pavilion building contrasts and compliments the monumental scale of the Millennium Dome, as illustrated in Figure 1.3.

Examples of the order of scale are as follows:

1. Volumetric scale: The big picture for the whole project and its locality.
2. Structural scale: The structural system, e.g. a 40 m span roof structure.
3. Module scale: A column grid, say, of 9 m.
4. Elemental scale: Repetitive elements, such as beams.
5. Assembly scale: The form of the connections.
6. Detail scale: The detail of the base of a column, or part of a truss.
7. Textural scale: Surface appearance.
8. Point scale: For example, the head of an individual bolt on a plate.

All of these elements of scale represent opportunities for architectural expression.

1.3 Pavilion at the Millennium Dome, Greenwich, UK (architect: Richard Rogers Partnership)

1.4.2 Ornament

In architectural composition, ornament has traditionally been sought in those places where portions of the building change significantly from one part to another, whether it be from wall to roof, wall to ceiling, one structural element to another, i.e. beam to column, or column to ground, and so on (see Colour Plates 9 and 27).

Much of the ornament and articulation of parts established in twentieth-century architecture has been found in the attention to the junctions between prefabricated components, whether they be parts of the structure or of the cladding systems. Consequently, in steel-framed buildings where the structure is exposed, ornament is usually sought in the connections between structural members and between the elements which comprise them (see Colour Plate 19). The careful shaping of the connection plates, stiffening elements, bolting and welding patterns, hubs for diagonal bracing and tie-rod assemblies, have taken on an important role, which is not only structural but also gives expression to the functionality of construction.

Examples where attention to detail can be used to provide ornamentation to an otherwise plain structure are:

- articulated attachment of horizontal and vertical members (Figure 1.2 and Colour Plates 12, 25 and 26)
- supports to arched members (i.e. at foundations) (Figures 1.3 and 1.4)
- suspension and bracing members, including tie rods (Figure 1.5 and Colour Plate 11)
- tie members that counterbalance a long-span portal frame (Figure 1.6)
- connections within trusses (Figure 1.7 and Colour Plate 20)
- fabricated beams or stiffened members (Figure 1.8 and Colour Plate 19)

1.4 Thames Valley University, pin-jointed connections supporting curved-arched steel members (architect: Richard Rogers Partnership)

1.5 Dynamic Earth Centre, Edinburgh (architect: Michael Hopkins & Partners)

1.6 St Paul's Girls School (architect: FaulknerBrowns)

1.7 Inland Revenue Headquarters, Nottingham, showing truss details which provide interest and articulation (architect: Michael Hopkins and Partners)

1.8 Operations Centre at Waterloo, London, showing a fabricated cantilever beam supporting a walkway (architect: Nicholas Grimshaw & Partners)

1.9 Orange Operational Facility, Darlington (architect: Nicholas Grimshaw & Partners)

1.10 Support to fabric roof at the Imagination Building, London (architect: Ron Herron and Partners)

- mullions with multiple perforations (Figure 1.9)
- support to a fabric roof (Figure 1.10 and Colour Plate 19).

Relationships can be established between the individual parts and the overall building form, which have a basis in elementary structural action.

1.5 Steel 'kit of parts'

A wide range of steel components is available to the architect and designer from:

- hot-rolled sections, such as I, H and L shapes
- tubular sections of circular, square and rectangular shape
- fabricated sections made by welding
- light steel components made from strip steel
- stainless steel components
- modular units made from light steel components.

This book concentrates on the application of hot-rolled and tubular-steel structures, but the principles are applicable to a 'kit of parts' of steel components. Indeed, in many buildings, I-sections are used for beams, H-sections for columns, tubular sections for bracing, fabricated sections for primary beams or transfer beams supporting columns, light steel for infill walls, and modular elements for plant rooms and toilets. Site connections are usually made by bolting, although welded connections may be preferred for factory-made connections.

More information on the technical characteristics of steel is presented in Chapter 11.

1.5.1 Hot-rolled steel sections

A wide range of standard steel sections is produced by hot rolling, from which designers can select the profile, size and weight appropriate to the particular application. Table 1.1 illustrates the range of 'open' sections used in the UK, which are universal beam (UB), universal column (UC), parallel flange channel (PFC) and angle sections.

In continental Europe, IPE and HE sections are generally used rather than UB and UC sections. In the USA, W and WF sections are used (which are similar to UB and UC sections).

Modern steel sections have parallel flanges. Also, parallel flange channels largely replace channels with tapered flanges. The 'serial size' refers to the designated section depth and width in which there are a range of weights of section. However, it should be noted that the UB or UC section designation, such as 406 × 140 × 39 kg/m refers to the *approximate* depth, width and weight, and exact dimensions should be obtained from standard tables.

Table 1.1 Standard hot-rolled sections (UB, UC, [and L)

Universal beam (UB)

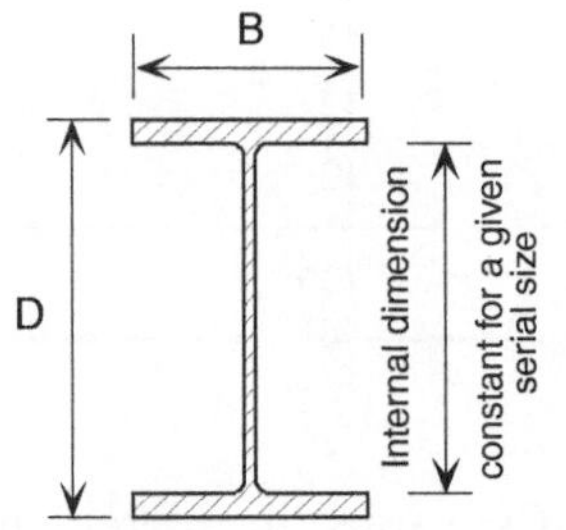

Universal Beams

Nominal dimensions (mm)

D	B
203	102 and 133
254	102 and 146
305	102 and 165
356	127 and 171
406	140 and 178
457	152 and 191
533	210
610	229 and 305

Deeper and shallower UB sections are available but are not listed here

Universal column (UC)

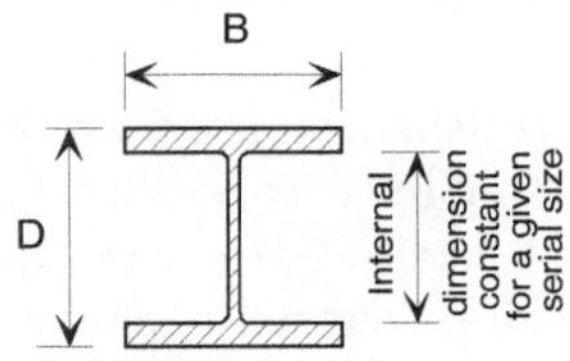

Universal Columns

D	B
152	152
203	203
254	254
305	305
356	368 and 406

Channel (PFC)

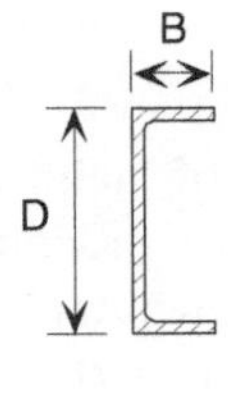

Parallel Flange Channel

D	B
100	50
125	65
150	75 and 90
180	75 and 90
200	75 and 90
230	75 and 90
260	75 and 90
300	90 and 100
380	100
430	100

Equal angle

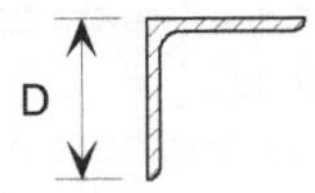

Rolled Steel Angles

External dimension of equal angle:

25, 30, 40, 45, 50
60, 70, 80, 90, 100
120, 150, 200, 250

Unequal angle

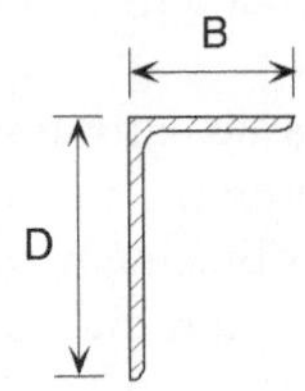

Rolled Steel Angles

External dimension of unequal angle

Various sizes from:
40 × 25 to 200 × 150

Including common sizes of:
75 × 50, 100 × 75, 150 × 75, 200 × 100

Table 1.2 Typical proportion of cost and man-hours per tonne in steel fabrication in buildings

Item	Man-hours per tonne	Total % of cost
Materials production	3–4	30
Fabrication	8–12	45
Erection	2–4	15
Protective treatment	1–2	10
Total	14–22	100

The serial size of UBs and UCs varies in increments of approximately 50 mm depth for the shallower sections, and 75 mm for the deeper sections. Within each serial size, the designer may choose from a number of different sections of similar height. The standardisation of steel sections has also led to the adoption of standard connections, which have become familiar within the industry.[5,6]

Table 1.2 shows the breakdown of costs in a typical framework of a building. It is apparent that only 30% of the cost associated with a steel frame relates to the material itself. Costs can increase significantly if fabrication and detailing create a demand for increased labour time. For example, using a heavier steel section is generally cheaper than using a lighter section that has to be stiffened at its connections.

1.6 Tubular steelwork

The use of tubular steelwork creates a wide range of architectural opportunities in internal or external applications. The word 'tubular' has come to mean applications using all forms of structural hollow sections, rather than just circular sections. Tubular sections are available as circular hollow sections (CHS), and square or rectangular hollow sections (SHS and RHS, respectively). Oval tubes are also available. SHS can also be used as the generic title 'structural hollow sections'. More detail on the methods of manufacture is presented in Section 11.3.

Table 1.3 defines the common section sizes. All tubular sections have *exact* external dimensions for detailing purposes.

The factors that influence the use of tubular construction are their:

- aesthetic appeal, which is often due to their apparent lightness of the members
- reduced weight of steel due to their structural efficiency, depending on their application
- torsional resistance (hollow sections are particularly good at resisting torsional effects due to eccentric loading)
- compression resistance for use as columns or bracing members (tubular sections are very efficient in compression due to their reduced slenderness in buckling conditions)

Table 1.3 Structural hollow sections (note, external dimensions are constant for a given serial size in all hollow sections)

	**Square Hollow Section (SHS)** External dimension of square section (mm): 40, 50, 60, 70, 80, 90 100, 120, 140, 150, 160 200, 250, 300, 350, 400
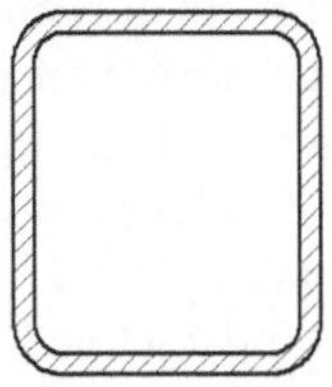	**Rectangular Hollow Section (RHS)** External dimension of rectangular section: Depth (mm) × width (mm). Various sizes from: 50 × 25 to 500 × 300 Including common sizes of: 100 × 50, 150 × 100, 200 × 100, 250 × 100 250 × 150, 300 × 200, 400 × 200, 450 × 250
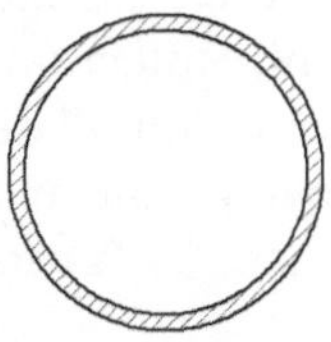	**Circular Hollow Section (CHS)** Size of CHSs Various sizes from: 21.3 mm diameter to 508 mm diameter Including common sizes of: 114.3, 168.3, 193.7, 219.1, 244.5, 273, 323.9, 355.6, 406.4, 457

- bending resistance of slender sections (if a beam is unrestrained throughout its length, the tubular section can be more efficient than a conventional I-section)
- efficiency under combined bending and torsion, such as in structures curved on plan
- fire and corrosion protection costs (which are reduced because of the low surface area of the tubular section)
- ease of site assembly, as also influenced by requirements for welding
- availability in higher grade S355 steel.

Fabricators cost all the steel-related items accurately, but the cost of fire and corrosion protection would normally be estimated separately. Some fabricators are specialists in tubular construction and can advise on costs and details at the planning stage. Additional aspects, such as the grinding of welds and special connection details, should be identified at this stage.

When using larger CHS, for example in long-span trusses, it is important to identify fabricators with specialist profiling equipment who can make the connections between the chords and web-members efficiently. This is particularly important for more complex assemblies, such as triangular lattice girders, which require a greater amount of fabrication effort and skill (see Section 6.8). The alternative may be to use SHS, which only require cutting the ends of the chord members at the correct angle, rather than profiling the cut ends.

The Waterloo International Terminal by Nicholas Grimshaw and Partners gains most of its visual impact by its striking lightweight roof. The roof consists of a series of tubular trusses supporting stainless steel cladding and glazing. Every truss is different, but considerable economy and simplification was achieved by repetition of the same external dimensions of the tubular sections (see Figure 1.11).

The excellent torsional resistance and stiffness of tubular sections (often ten times greater than that of I-sections of equivalent area), makes them suitable for curved bridges and canopies where members curve on plan and possibly also on elevation (see Colour Plate 14). Architects such as Santiago Calatrava have utilised this property by creating tubular spine-beams and inclined arches that resist eccentric loading in bending and torsion. An excellent example of the use of tubular-inclined arches is in the Millennium Bridge, Gateshead (Colour Plate 16).

Transportation buildings have also exploited the qualities of tubular construction. Examples include Stansted and Stuttgart Airports (see Figure 1.12 and Colour Plate 9, and also see Colour Plate 7). One of the largest buildings in the world employing tubular steel is the International Airport at Kansai, Japan, designed by Renzo Piano (see Figure 1.13).

Tubular structures are not only reserved for large projects. The lightness of tubular members is emphasised at the 'Gateway' in Peckham, London (see Colour Plate 10). Similarly, the inclined tubular members created a curved appearance in Hodder Associates' enclosed pedestrian footbridge in Manchester (see Colour Plate 17).

1.11 Waterloo International Terminal, striking long-span roof comprised of tapering tubular trusses (architect: Nicholas Grimshaw & Partners)

1.12 Stuttgart Airport Roof using tubular column 'trees' — see also Colour Plate 9 (architect: Von Gerkan Marg & Partners)

1.13 Tubular trusses at Kansai Airport, Japan (architect: Renzo Piano Workshop)

1.6.1 Fabricated sections

Fabricated steel sections are produced by welding steel plates in a factory process. These sections are fabricated to the required geometry and are not standard sections. They are usually economic where:

- the section size can be 'tailor-made' to the particular application and member depth
- long-span primary beams would not be achievable using hot-rolled sections
- heavy 'transfer' or podium structures are required to support columns or other heavy loads
- asymmetric sections are more efficient than standard sections
- tapered sections are specified, e.g. in grandstand canopies
- curved members are created by cutting the web and bending the flange into a curve of the required radius.

The use of fabricated sections in a floor grillage is presented in Section 4.2. The primary practical consideration is the availability of standard plate sizes and the relative thicknesses of the plates used in the flanges and web of the section. Examples of the use of fabricated members are also illustrated in Section 2.4.

1.6.2 Cold-formed sections

A variety of cold-formed sections (CFSs) are produced and these sections are widely used as secondary members, such as purlins, or in light steel framing for primary structural applications. Typical C sections are illustrated in Table 1.4. CFSs are produced by cold rolling from galvanized strip steel in thicknesses of 1.2 to 3.2 mm for structural applications. Various SCI publications, including an

Table 1.4 Cold-formed sections (produced by various manufacturers)

Standard C and Z sections 	Depth of C or Z sections (typical) (mm): 75, 100, 125, 140, 170 200, 240, 300
Modified C and Z sections 	Steel thickness (typical) (mm): 1.2, 1.6, 1.8, 2.0, 2.4, 3.2

Architect's Guide,[6] describe the use of cold-formed steel sections and light steel framing in building.

Cold-formed steel sections can be used as:

- infill or separating walls in steel framed buildings
- floor joists and secondary members in frames
- light steel framing, erected as storey high wall-panels, in housing and residential buildings
- purlins in roofs and in over-roofing in building renovation
- modular units in cellular building forms, such as hotels and student residences
- cladding support members and over-cladding in building renovation.

Steel decking is produced in steel thicknesses from 0.9 to 1.25 mm, and is available in two generic forms:

- deck profiles of 45 to 80 mm depth for use in composite construction
- deep deck profiles of 210 or 225 mm depth for use in *Slimdek* construction.

These applications in composite construction are covered in more detail in Section 4.1.

1.6.3 Modular construction

Modular construction uses prefabricated 'volumetric' components which are generally made from light steel-frames, although they often incorporate SHS columns for the corner 'posts'. It is most economic where the modules can be manufactured repetitively to achieve 'economy of scale', and where the dimensions of the modules are suitable for transportation and installation (3.0 to 4.2 m are typical module widths).

The Peabody Trust's Murray Grove project in London achieved architectural acclaim by being the first major use of modular

1.14 Modular construction of social housing in London for the Peabody Trust (architect: Cartwright Pickard Architects)

construction in the social-housing sector (see Figure 1.14). The modular nature of the building was softened by using prefabricated balconies, access walkways, and a 'core' lift and stairs structure at the axis of the two wings of the building.

More guidance on the use of modular construction can be found in recent SCI publications.[7,8] Modular units can be used in more regular framed structures in the form of prefabricated plant rooms, 'clean rooms', bathrooms and toilets, which are often lifted or slid into place on the floor.

Chapter 2

Introduction to expressed structural form

The visual expression of a structure requires an understanding of structural function, and an appreciation of the alternative forms of structure that can perform this function. Broadly, the various forms of steel structure that may be encountered may be grouped as follows:

- Braced frames, in which the beams and columns are designed to resist vertical loads only. Horizontal loads are resisted by bracing in the walls or cores. The connections are designed as pinned or 'simple'.
- Rigid or 'sway' frames, in which the framed structure is designed to resist both vertical and horizontal loads by designing the connections between the members as moment-resisting.
- Arch structures, in which forces are transferred to the ground mainly by compression within the structure.
- Tension structures, in which forces are transferred to the ground by tension (or catenary action) and by compression in posts or masts, as in a tent. The tension elements in the form of cables or rods are usually anchored to the ground.

These structural systems are explained in more detail in Chapters 3 and 7. Tension structures are commonly associated with expressive external structures. In practice, many structures are a hybrid of two or more forms. For example, a portal frame acts as a rigid frame in one direction, but is braced in the other direction.

Pinned connections are usually simple to fabricate and are the least expensive type of connection to produce. In pin-jointed frames, lateral stiffness must be introduced into the frame by the careful placement of diagonal bracing, or by incorporating other stabilising methods, such as shear walls, stiff cores, or by interaction with rigid frames. Of course, pin joints do not have to take the form of 'pins'. Rather, they are simple connections that are treated as pins from the point of view of structural design. Actual pins may be treated ornamentally, as shown in Figure 2.1, and as used at the Sackler Gallery in London (Figure 2.2). Often, connections are designed as 'pinned', even if they possess some rotational stiffness.

The notion of a rigid frame relates to the stiffness of the connections rather than to the rigidity of the frame itself. The

2.1 Pin-jointed connections to column, sports centre, Buchholz, Switzerland (architect: Camerzindgrafen Steiner)

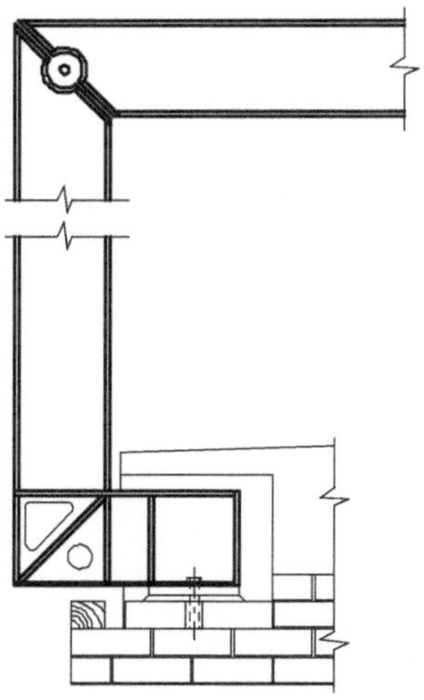

(a)

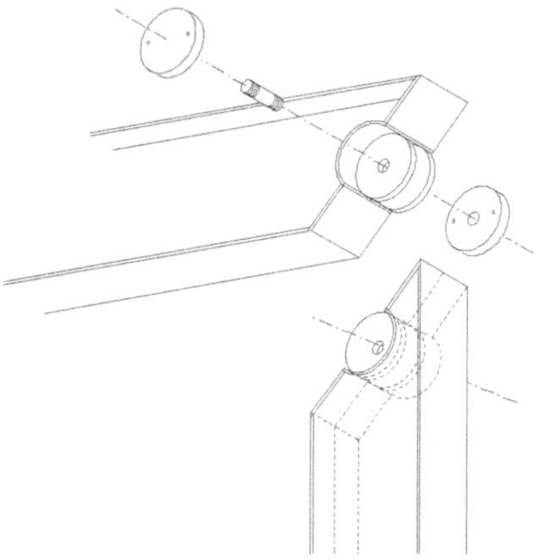

(b)

2.2 Pinned connection between column and beam at the Sackler Gallery, London: (a) column–beam arrangement; and (b) local detail (architect: Foster and Partners)

achievement of full continuity between members in rigid frames requires an extensive amount of fabrication and, as a consequence, is relatively expensive to achieve. However, rigid frames are suitable for low-rise buildings and enclosures, where horizontal forces are low in relation to vertical loads.

Given the overall geometry of any one particular structural arrangement, there are many different types of connection which can be made between the members. However, the selection of the structural members, both in their own cross-section and in their connection to other members, must be known before any structural analysis is carried out. There is therefore a close inter-relationship between the architectural requirements for choice of the frame members and their detailed structural design.

Often, rigidly framed structures are preferred if there is little opportunity for the use of vertical bracing, such as in fully glazed façades or in large-span structures.

Arch and tension structures rely on the compressive and tensile properties of steel, and follow well-defined structural principles (see Section 2.2 and 2.3).

2.1 Expression of bracing

Of the several methods used to achieve lateral stability in framed construction, diagonal bracing is the one which offers the clearest and most direct visual and graphic representation. For this reason, bracing has been used as an explicit form of structural expression. When brought to the exterior, bracing is often used to ornament the building as well as to serve a structural function. Bracing used for compositional effect can be more than the minimum necessary for structural purposes, as is the case in Figure 2.3.

Figure 2.4 shows a new visitor centre for a thirteenth-century castle, which was built over some of the archaeological relics. The structure had to be lightweight to reduce the size of the foundations, and the number of columns had to be limited to avoid interference with the exhibition space below. The inherently lightweight nature of the building is expressed by an external structure, with the diagonal bracing adding a further element of interest.

Often, the location and orientation of the bracing has to satisfy other criteria, such as the provision of large openings or the spatial alignment of the cladding elements. In this case, the range of architectural options for bracing systems is constrained by the building function.

In multi-storey buildings, bracing can be expressed externally to architectural effect. The structural importance of the bracing members means that their size and detailing must conform to sensible load paths by minimising eccentricities and points of weakness. Welded stiffeners are often required to transfer forces across highly stressed members.

2.3 Reliance Controls, Swindon, with multiple braced panels (the stubs of the steel roof beams were left exposed to facilitate easy extensions to the factory) (architect: Foster and Partners)

2.4 Visitor centre, Limerick, Ireland. The structure of the building is expressed on the outside, including the vertical bracing (architect: Murray O'Laoire Associates)

2.2 Arched and curved structures

Arches are convex structures that are designed primarily to resist compression, as a result of their shape and the form of loading acting on them. Arches are theoretically of parabolic form when subject to uniform loading, but they can be circular, or even made from multiple linear elements. Arches also resist bending moments which are also induced due to non-uniform loading, or the deviation of the arch from the idealised shape in which the lines of thrust (compression) are located within the member cross-section.

Arches in steel may be made of I-sections that are either curved to shape (see Colour Plate 22), or made as a facetted arch from multiple straight lengths. They can also be in the form of fabricated members, such as trusses. Arches may have rigid or pinned bases, or a combination of both. Figure 2.5 shows an excellent example of external and internal arches within a multi-storey building used to great structural advantage by spanning over the railway lines at the Broadgate development, London.

Tubular members are excellent for use in arch construction because of their resistance to buckling and, hence, the few lateral restraints that are required.[9] At the Windsor Leisure Centre, an arch with variable curvature was continued outside the building envelope, as shown in Figure 2.6. See curved tubular trusses in Colour Plate 4 and the glazing supports in Colour Plate 14.

The roof of the great glasshouse of the National Botanical Garden of Wales used the concept of a curved roof consisting of arches of similar curvature but of reducing span to create a toroidal shape (like a slice through a car tyre), as illustrated in Colour Plate 2. The maximum span of 60 m is achieved with only 324 mm diameter circular hollow sections (CHSs) which support the glass roof.

Hong Kong's new airport uses a variety of novel construction forms, including a long curved canopy over the walkway, as illustrated in Figure 2.7.

Steel members may also be curved in the horizontal plane rather than in the vertical plane, as illustrated in Figure 2.8. In this case, the

2.5 Internal arch structure over railway lines at Broadgate, London (architect: Skidmore Owings & Merrill)

2.6 Arched roof at Windsor Leisure Centre (architect: FaulknerBrowns)

2.7 Curved members at Hong Kong International Airport (architect: Foster and Partners)

2.8 Curved canopy at the Strasbourg Parliament (architect: Richard Rogers Partnership)

members are subject to bending and torsion, which is a complex interaction. Inclined curved members can also be used to great effect but, in this case, additional horizontal or torsional support is required to counterbalance the forces. The Merchant's bridge in Manchester utilises this principle, as illustrated in Colour Plate 18.

2.3 Tension structures

In tension structures, 'ties' are designed to resist only tension and are crucial elements in the overall structural concept. Tall compression members or 'masts' provide for the necessary vertical support, and these masts are located fully or partially outside the enclosure. Cable-stayed roofs, suspended structures, cable nets and membrane structures are all types of tension structures. Good examples of this form of construction are shown in Figures 2.9 and 2.10.

2.9 Homebase, London, showing a central spine supported by tension rods from mast (architect: Nicholas Grimshaw & Partners)

2.10 Tension-supported membrane roof to a central amenity building, UK side of the Channel Tunnel (architects: BDP)

Tension structures can have clear advantages for the roofs of long-span structures or enclosures where the internal function of the space is crucial. In smaller-scale applications, their use is more likely to depend upon a combination of technical and architectural arguments, such as the desire for a lightweight or membrane roof, or to support a glass wall with the minimum of obstruction.

The tension structure may be partially or fully exposed, and both appearance and function are equally important to detailing. Common examples of structures where tension elements act as primary members include:

- roofs for sports buildings, halls and auditoria
- grandstands (Colour Plates 24 and 25)
- canopies
- cable-stayed bridges and walkways
- some 'high profile' buildings and structures (Colour Plate 23)
- railway, airport and other buildings for transportation
- tall glazed walls.

2.11 Tied column used for architectural effect at Stockley Park, London (architect: Foster and Partners)

At a modest scale, the tapered columns and cantilevered arms of the roof structure at Stockley Park, London, are supported by horizontal ties, which also enhance the visual effect, as in Figure 2.11.

The details employed in tension structures are covered in Chapter 7. Tension forces are resisted externally by concrete foundations or by tension piles. The economic design of these components is also important to the overall concept.

2.4 Fabricated members

Fabricated steel sections can be made of a variety of components:

- steel plates to create I-beams, or tapered beams (see Colour Plates 26 and 27)
- I-beams cut into Tee sections (see Colour Plate 19)
- tubular sections with welded fins.

The following figures illustrate the wide range of architectural effects that can be created by fabricated sections, often at large scale.

The Financial Times building in Docklands, London, used half tubes, welded plates and projecting fins to create wing-shaped columns that are external to the fully glazed envelope. The overall effect is illustrated in Figure 2.12, and the cross-section of the column is shown in Figure 2.13.

The curved beams at Stratford Station, London, were welded from plate and stiffened at points of high curvature, as illustrated in Figure 2.14. Interestingly, cast steel footings connected the curved beams to the concrete ground beam, and accentuated the local curvature.

The enclosure of BCE Place in Toronto, based on a concept of Santiago Calatarva, used tapered columns comprising four fins that reduce to a fine arch over the enclosure between adjacent buildings (Colour Plate 8).

The Cranfield Library used a welded V-shaped spine-beam to support the curved roof, as illustrated in Figure 2.15 and in the detail in Figure 2.16.

2.12 Financial Times building, London (architect: Nicholas Grimshaw & Partners)

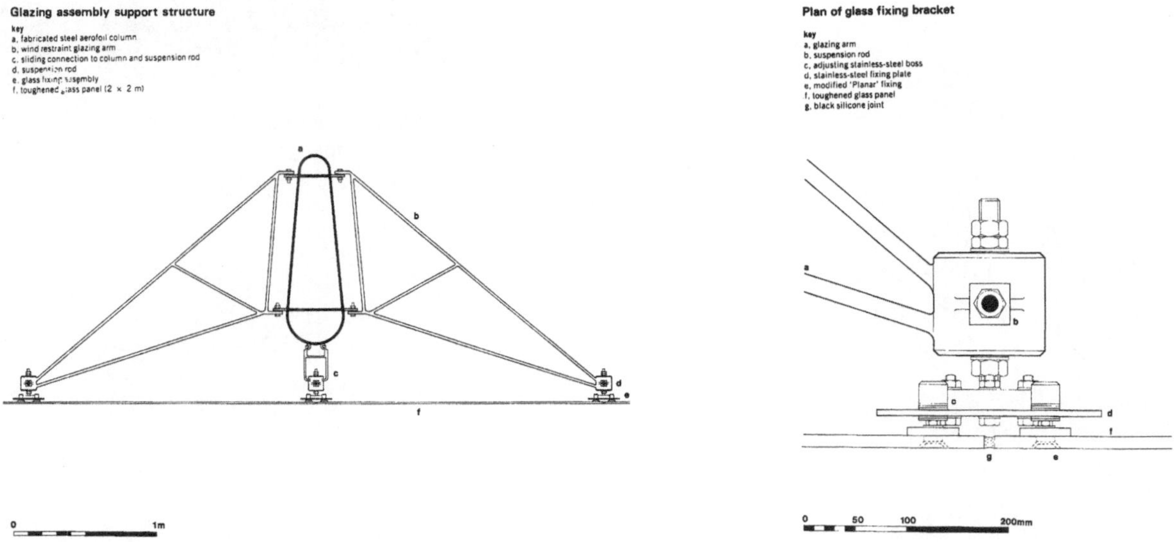

2.13 Details of column in Figure 2.12, Financial Times building, London

2.14 Curved fabricated beam at Stratford Station, London (architect: Wilkinson Eyre)

2.15 Fabricated V beam of Cranfield Library (architect: Foster and Partners)

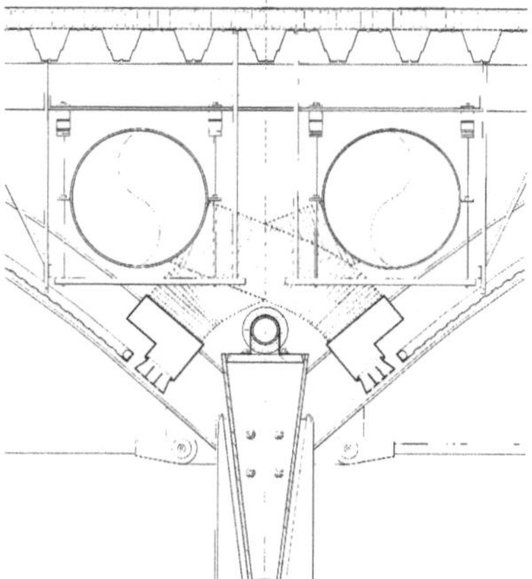

2.16 Detail of a fabricated section used at Cranfield Library

2.5 Structure/envelope relationship

Steel is often used in applications in which the relationship with the building envelope is important to the visual effect. There are five basic relationships between the enclosure of a building and the primary structure:

- structure located entirely inside the building envelope (Colour Plate 1)
- structure located in the plane of the building envelope (Colour Plate 12)
- internal structure continued outside the building envelope (Figures 1.1, 1.11 and Colour Plate 13)
- semi-independent external structure supporting external wall, glazing or roof (Colour Plate 13)
- structure located completely outside the building envelope (Figure 2.17 and Colour Plate 6).

At Bedfont Lakes, London, the beams and columns were located in the plane of the building envelope and used expressed connections, as illustrated in Colour Plate 12. Many 'tent-type' structures continue to support the structure through the building envelope, as was done at the Dynamic Earth Centre, Edinburgh, illustrated in Figure 1.5. The same concept was also used in the glazed cladding support to the Western Morning News building, Plymouth, illustrated in Colour Plate 13. An early example of a completely external structure is the Inmos factory in Newport, south Wales, shown in Figure 2.17.

2.17 Structure outside the building envelope, Inmos, Newport (architect: Richard Rogers Partnership)

2.18 Stratford Station showing use of repetitive curved frames (architect: Wilkinson Eyre)

This relationship between the envelope and the primary structure brings with it other issues important to the building design, such as:

- expression of the connections
- foundation and holding down points
- security and access (for external structures)
- fire-safety strategy
- corrosion protection of the external elements
- 'cold-bridging' through the envelope
- secondary supports to the roofs and walls to complement the chosen structural solution.

The mixture of structural elements, including curved members, trusses, fabricated components, cables, cast and stainless steel elements, illustrates the variety of techniques that are achievable.

The repetition of the internal structure externally can also be used to great visual effect. The curved frame of Stratford Station in Figure 2.18 was extended outside the glazed façade to emphasise the structural solution. The tapered fabricated beams were curved with decreasing radius down to heavy cast steel footings. This same notion was first used in the Centre Pompidou, Paris, where the external framework replicates the internal structure.

Buildings can be extended later by using the external framework to connect into the structure of the extended building without having to remove the existing cladding. This is important in the operation of existing buildings, which would otherwise lead to disruption of internal activities.

Chapter 3

Frame design

3.1 The frame as the basic unit of construction

A framework is a three-dimensional assembly of steel members that form a self-supporting structure or enclosure. The most common and economic way to enclose a space is to use a series of two-dimensional frames that are spaced at equal intervals along one axis of the building, as shown in Figure 3.1(a). Stability is achieved in the two directions by the use of rigid framing, diagonal bracing, or through the supporting action of concrete shear walls or cores (see Section 3.2). This method of 'extruding' a building volume is equally applicable to any frame geometry, whether of single or multiple bays.

Three-dimensional frames can vary enormously in overall form, in the overall geometry of the individual members comprising them, and in the elements comprising the horizontal and vertical members. In these more complex frames, elements may be repeated, but the

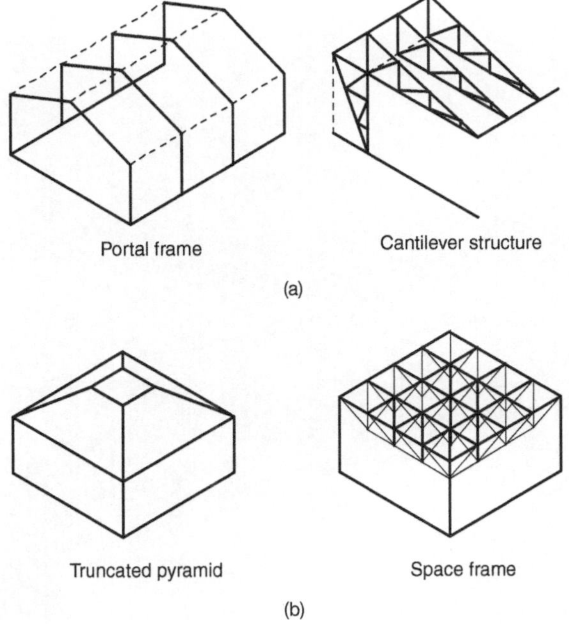

3.1 Examples of various forms of two- and three-dimensional frames to form enclosures: (a) two-dimensional frames (repeated to form a three-dimensional structure); and (b) three-dimensional frames (repeated parts relying on mutual support)

structure relies for its effectiveness on mutual support in three dimensions (see Figure 3.1(b)).

Multi-storey building frames comprise beams and columns, generally in an orthogonal arrangement. The grillage of members in the floor structure generally comprise secondary beams that support the floor slab and primary beams that support the secondary beams. The primary beams tend to be heavier and often deeper than the secondary beams. Various structural alternatives for these members are presented in Chapter 4.

3.2 Exposing the frame

The exposure of the frame, either in part or in whole, obviously depends upon the relationship between the skeleton and external skin. The frame can be located completely external to the cladding, in which case it is given expression in the external appearance of the building. Alternatively, the frame can be located wholly internal to the cladding, in which case it may find little or no expression externally. Between these two extremes, the interaction of the frame and cladding establishes a further range of relationships. Buildings of an entirely different character emerge depending on these spatial relationships.

A simple example of a portal-frame structure that is continued outside the building envelope to visual effect is shown in Figure 3.2. In this case, the perforated cellular beams enhance the lightness of the structure whilst preserving its primary function as a rigid frame.

Basic building physics requirements, in terms of thermal insulation and control of condensation, also have to be addressed, particularly when the frame penetrates the building fabric (see Chapter 10).

3.2 Portal-frame structure created using cellular beams

3.2.1 *Repetition of frames*

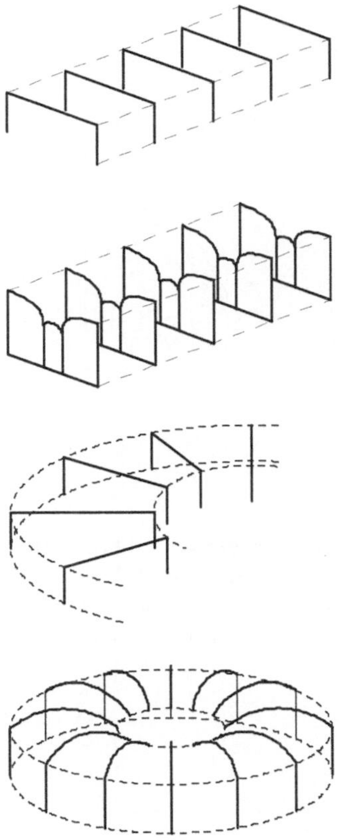

3.3 Various illustrations of identical frames repeated at intervals

An exposed structure establishes a dominant rhythm in the elevational composition. More often than not, it is a simple and singular rhythm derived from the equal spacing of the primary frames. Various examples of repeated frames to form larger enclosures with increasing complexity are shown in Figure 3.3.

An external framework or skeleton often demands greater attention to detail, but conversely permits greater freedom in choice of structure form, as the structure is no longer dependent on the spatial confines of the internal envelope. Therefore, tension structures find their true expression in external structures (refer to Chapter 7).

3.2.2 *External frames*

By selectively exposing or concealing structural members, emphasis can be given either to the primary frames, or to the wall and ceiling planes which define the building volume. In one of the early examples, Mies van der Rohe's Crown Hall building (see Figure 3.4), the large-span portal frame is clearly expressed, yet subtly woven into the fabric of the external wall. In other structures, a clearer distinction is made between the external frame and building enclosure, such as by use of masts and cables in tension structures.

The Lufthansa terminal at Hamburg Airport uses a portal frame comprising plated box-sections to create a massive external skeleton (Figure 3.5).

3.3 Braced versus rigid frames

The fundamental structural requirement governing the design of connections in building frames is related to the strength and stiffness

3.4 Crown Hall: external portal frame (architect: Mies van der Rohe)

3.5 Lufthansa Terminal, Hamburg Airport (architect: Von Gerkan Marg & Partners)

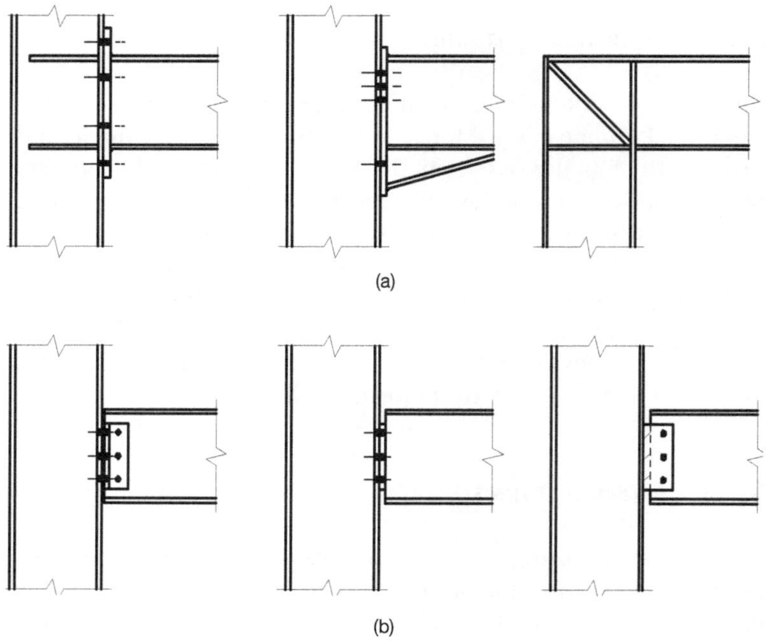

3.6 Various forms of steel connections: (a) examples of effectively 'rigid' connections; and (b) examples of effectively 'pinned' connections

of the connections between the members, or of members to the foundations.

The connections may be one of three configurations defining these degrees of strength (or more correctly 'resistance') and stiffness:

1. Rigid (also called fixed or moment-resisting) connections (Figure 3.6(a)).
2. Pinned (also called simple) connections (Figure 3.6(b)).
3. Semi-rigid (also termed partial strength) connections.

Rigid frames require rigid connections in order to provide for stability at least in one direction. Braced frames are stabilised by vertically oriented bracing, and require only pinned connections. Rigid frames are often termed 'sway frames', because they are more flexible under horizontal loads than braced frames.

The characteristics of these connections are presented in more detail in Chapter 5 and may be summarised as follows.

In a *'rigid' connection* there is complete structural continuity between any two adjacent members. Moment (or rigid) connections are used in frames where there is a desire to omit vertical bracing in one or both directions. The main advantage of rigid frames is that an open space between columns can be created, which offers flexibility in choice of cladding, etc. (e.g. in glazed façades). However, the achievement of full continuity between members at the connection requires an extensive amount of fabrication and, as a consequence, this system is relatively expensive.

To achieve a nominally *'pinned' joint*, the connections are made so as to permit the transfer of axial and shear forces, but not bending moments. Nominally simple connections may provide some small degree of rigidity, but this is ignored in structural design and these connections are treated as pinned. Examples of pinned connections are cleated, thin or partial depth end-plates, and fin-plate connections as illustrated in Figure 3.6(b).

Pinned connections are usually simple to fabricate and erect, and are the least expensive type of connection to produce. As a consequence, lateral stiffness must be introduced into the frame by other means.

Semi-rigid (and also partial strength) connections achieve some continuity through the connections, but are not classified as full strength, as they do not achieve the bending resistance of the connected members. These forms of connections are illustrated later on in Figure 5.5. They are used for low-rise frames in which horizontal forces are not so high, or in beams where some end fixity is beneficial to the control of deflections.

3.4 Portal-frame structures

Portal-frame type structures are examples of rigid frames that can take a number of forms. They were first developed in the 1960s, and have now become the most common form of enclosure for spans of 20 to 60 m. Portal frames are generally fabricated from hot-rolled sections, although they may be formed from lattice or fabricated girders. They are braced conventionally in the orthogonal direction.

In general, portal-frame structures are used in single-storey industrial type buildings where the main requirement is to achieve a large open area at ground level and, as such, these structures may not be of architectural significance. However, the basic principles can be used in a number of more interesting architectural applications, as illustrated in Figures 3.2 and 3.7. Also, portal frames can be used in other applications, such as in roof structures for multi-storey buildings, long-span exhibition halls, and atrium structures.

The frame members normally comprise rafters and columns with rigid connections between them. Tapered haunches are introduced to strengthen the rafters at the eaves and to form moment-resisting connections. Either pinned or fixed bases may be used. Roof and wall bracing is essential for the overall stability of the structure, especially

3.7 Portal frame expressed internally behind a glazed-end elevation of a building for Modern Art Glass (architect: Foster and Partners)

during erection. Typical examples of portal-frame structures using hot-rolled sections, fabricated sections and lattice trusses are illustrated in Figure 3.8. Portal frames generally provide little opportunity for expression but, with care, the chosen details can enlighten the appearance of these relatively commonplace structures.

Other applications of portalised structures are illustrated in Figures 3.9 and 3.10. The articulated lattice structure using tubular elements was used to great effect in the Sainsbury Centre, Norwich. An arch or mansard shape can be created from linear members, as in Figure 3.11.

In tied portals, the horizontal forces on the columns may be restrained by a tie at, or close to, the top of the column. Ties are usually not preferred because they can interfere with the headroom

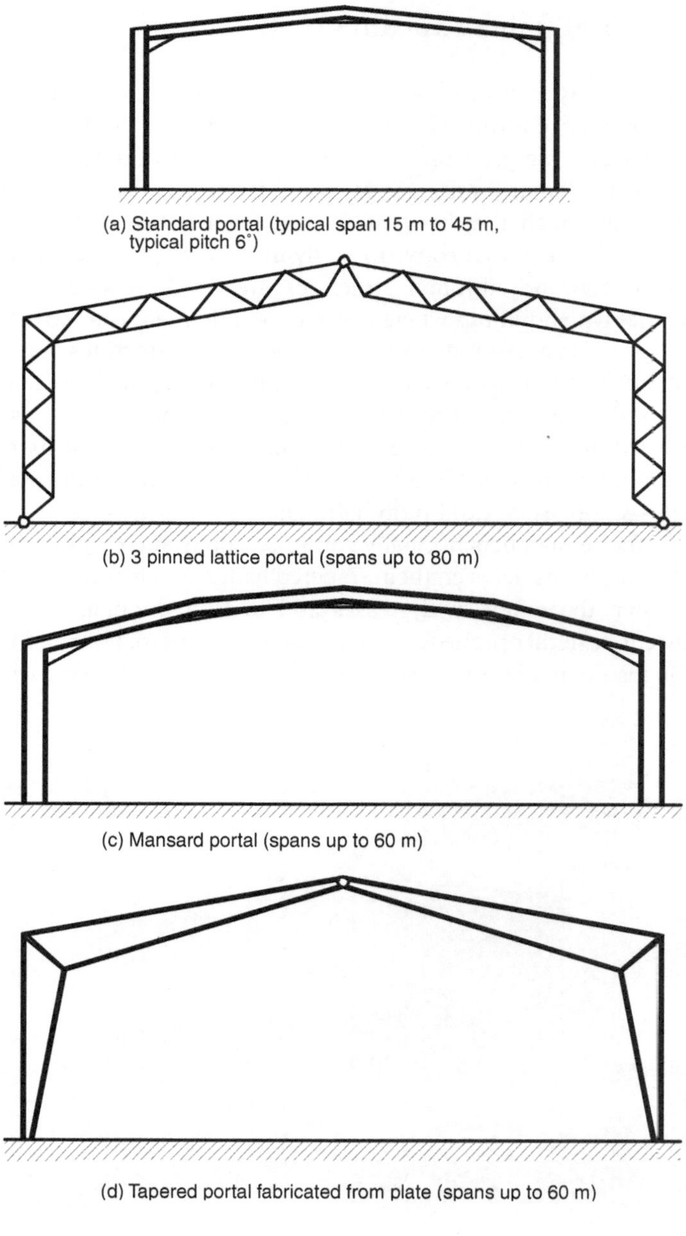

3.8 Typical portal structures using a variety of members

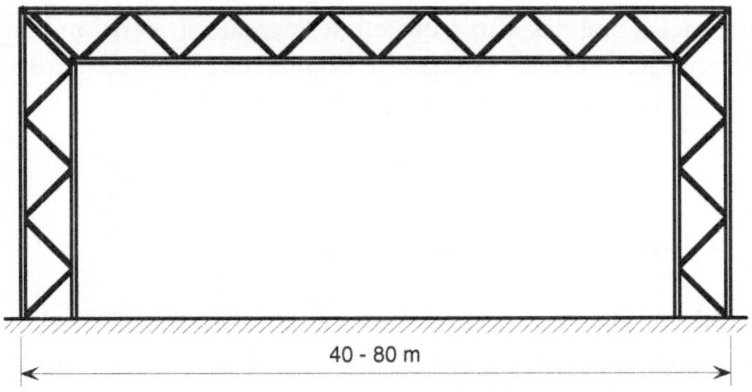

3.9 Articulated lattice portal structure (often using tubular sections)

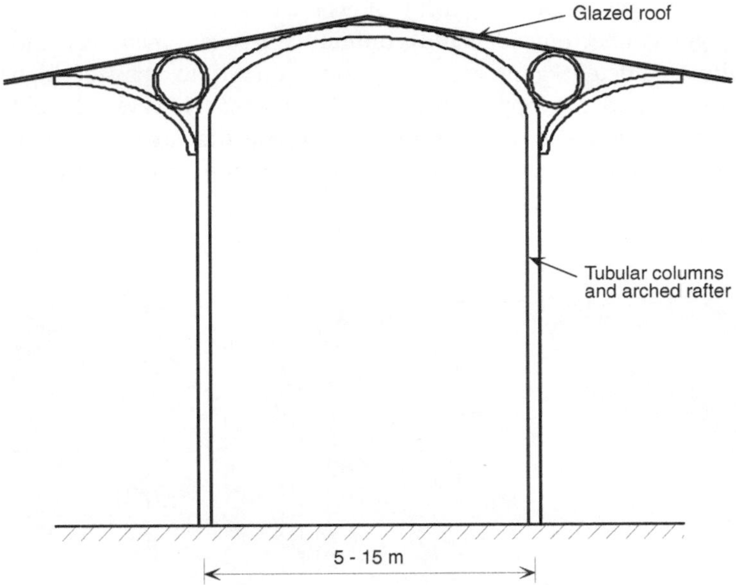

3.10 Arched portal using tubular sections

3.11 Long-span portal frame used to create an arch structure

3.12 Tied portal frame used at Clatterbridge Hospital (architect: Austin-Smith: Lord)

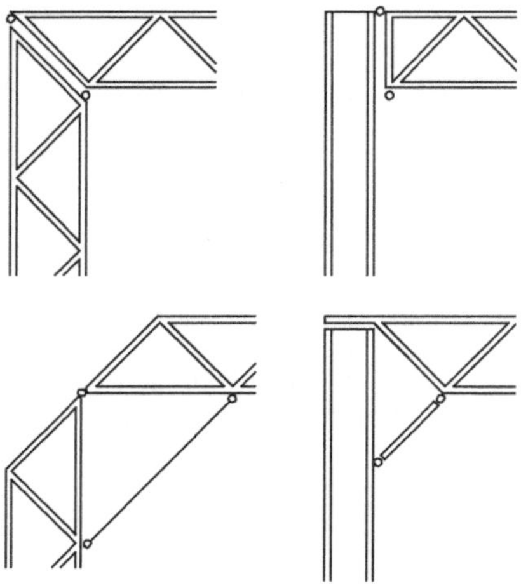

3.13 Rigid connections achieved by pinned connections between the elements

of the space. Long ties also require intermediate suspension support to prevent sag. However, ties can be detailed effectively, as illustrated in Clatterbridge Hospital in Figure 3.12.

3.5 Expressing the connections

Connections exert a strong influence on the architectural form. Pinned and rigid connections are quite distinct and produce quite different forms and details. The discontinuity of a pinned connection can either be accentuated and given a clear expression in the structural form, or, alternatively, it can be made less apparent. By drawing such distinctions in relation to the individual frame, and then to the whole building form, offers the basis for expression.

Rigid connections demand continuity between members and invite a different approach. They are required to transfer high moments and can appear heavy and complex. However, a rigid connection may also be achieved through parts that are pin-jointed, as simplified in Figure 3.13, and by example in the Sainsbury Centre

3.14 Example of continuity achieved through a series of pinned connections, Centre Pompidou, Paris (architect: Renzo Piano and Richard Rogers)

in Figure 1.1. In these cases, moments are transferred by tension and compression in the connections.

The end wall of the Centre Pompidou in Paris, shown in Figure 3.14, illustrates an unusual application of the principle, where the typical pinned connection between the truss and column is elegantly transformed to a moment-resisting connection by the addition of a continuous tie from the 'gerberette' extension to the truss and attached to the foundations.

Depending upon the exact nature and locations of connections in a frame, the 'reading' of the individual members and the frame as a whole can vary markedly. This is further illustrated in Figure 3.15 for a three-bay frame, in which different formal relationships between members and individual bays are established by simply varying the locations of the pinned connections in the structure. All cases are structurally admissible, but can create entirely different details.

A good example of articulation within a structure is illustrated in Figure 3.16. Inclined 'arms' support slightly curved rafters and create a portal frame effect, allowing the connections to be expressed as nodes.

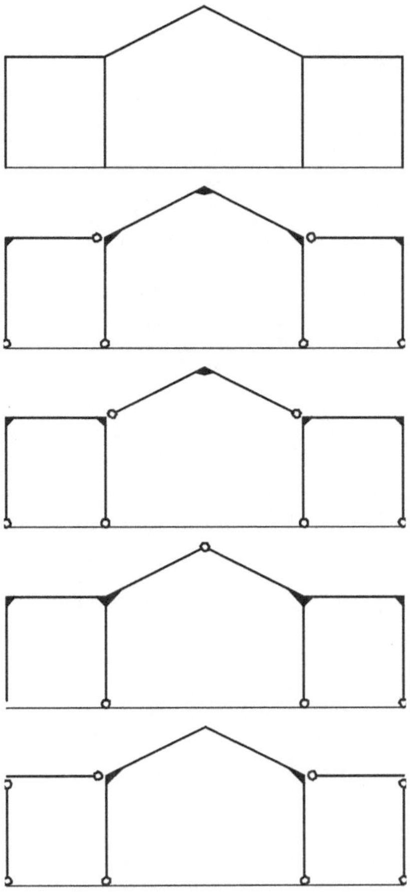

3.15 Different overall forms of the frame by varying type and location of pinned connections

3.6 Alternative forms of bracing

Nominally pin-jointed frames are braced in the vertical and horizontal directions. 'Braced' structures can be achieved in a variety of ways, including full-height bracing of a bay between columns, or a shorted 'knee' bracing to achieve hybrid action between a braced and a sway frame (as illustrated in Figure 3.17).

Often, the floor structure can act as 'plan' or horizontal bracing, but in single-storey buildings, separate horizontal bracing is required in the plane of the roof to transfer loads to the vertical bracing in the walls or cores.

3.6.1 Vertical bracing

The stability of the building is dependent on the form and location of the vertical bracing, or other shear-resisting elements which are linked by floors or horizontal bracing.

For simplicity, vertical bracing is located in the façade or internal separating walls. Ideally, the bracing line would be on the centre-line of the main columns, but this may conflict with the location of the inner skin of external walls. Discussion between the architect and the structural engineer at an early stage can resolve this difficulty. Often, flat steel bracing elements are located in the cavity of the masonry wall to minimise these dimensional problems.

The most common arrangements of bracing in multi-storey construction is 'X', 'V' or 'K' bracing using steel angle or circular hollow sections (see Figure 3.18). Inverted 'V' bracing is preferred where substantial openings, e.g. doors, are required in the braced bay. To reduce its visual impact, bracing is often positioned around

3.16 Portal-frame effect created using inclined pinned members

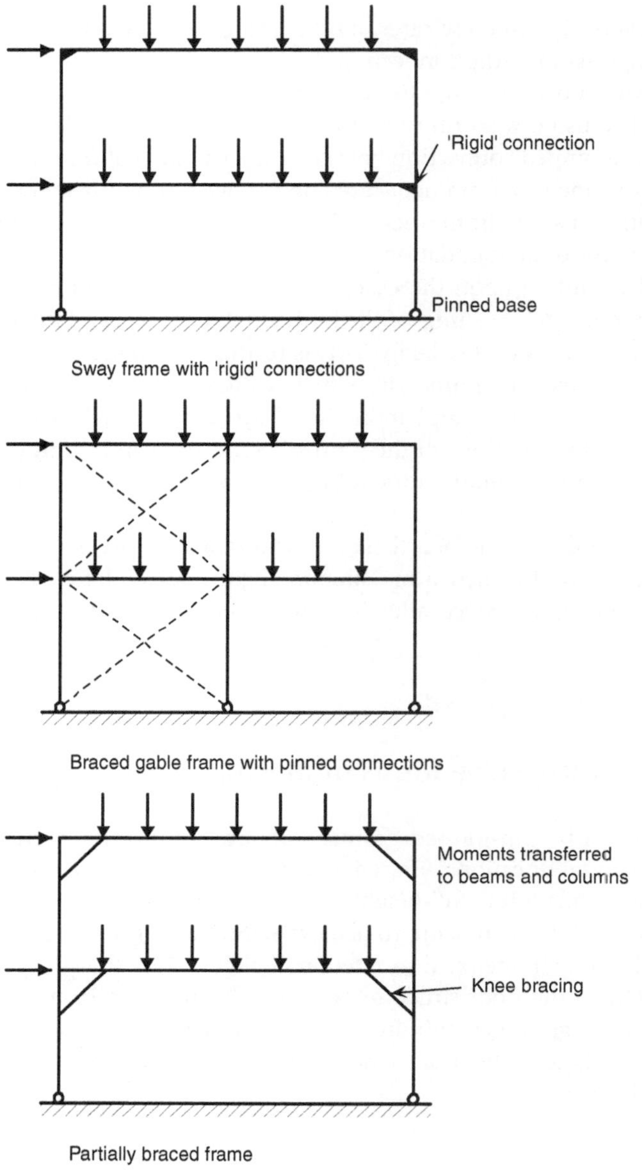

3.17 Examples of rigid and braced frames

vertical cores, which usually house the lifts, stairs, vertical service ducts and/or toilets, or on the external face of the building within the cavity wall.

Figure 3.18 also illustrates the forces in the individual members. In the X-braced form, the members may be designed to resist both tension and compression, or tension only, which leads to more slender members. Tension rods or flat plates are largely ineffective in compression, and, therefore, forces are resisted only in tension when using these elements. In the K- and V-braced forms, the members must be designed to resist tension and compression, depending on the direction of the forces on the building. Tension ties are not possible in this case.

Tension tie members are generally used in exposed steelwork because of their apparent 'lightness'. In X-braced frames, special

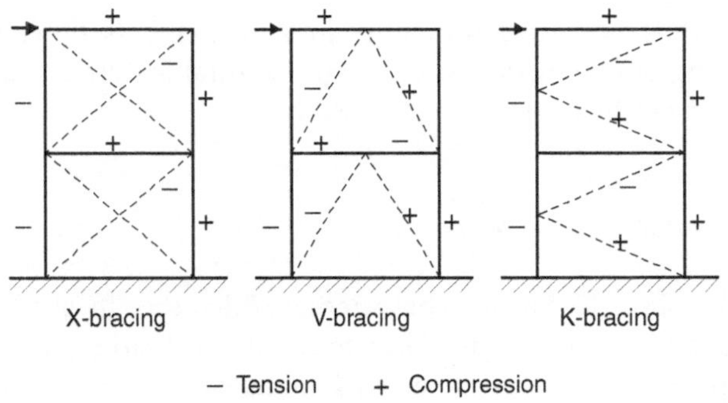

3.18 Different forms of bracing and their forces

brackets may be included to allow connection of the four tie members at the cross-over points. An example of an X-braced structure using CHS sections with a connecting plate is illustrated in Figure 3.19.

A 'hybrid' between a rigid frame and a braced frame can be achieved by the use of 'knee' bracing. In this case, the corner junction between a beam and column is stiffened by a short bracing member, which is designed to resist either tension or compression (see Figure 3.17). The bracing member transmits a force to the beam or

3.19 X-bracing using CHS sections used at a sports centre in Hampshire (architect: Hampshire County Council)

columns, which is resisted by bending in these members. If necessary, knee bracing can be expressed as an architectural feature by curving the members or by using cast inset pieces.

3.6.2 *Concrete or steel cores*

As an alternative to bracing the external walls, the lift shafts and stairwells can be used as rigid 'cores' to stabilise the structure. Braced or steel-plated cores can be erected along with the rest of the steelwork, whereas concrete cores are generally built in advance of the frame and can be slower to construct. Accuracy is required for the installation of lift guide rails,[3] which is affected by the verticality and accuracy of the cores. Furthermore, multiple openings for service penetrations and doors can affect the stabilising effect of the core. It is not unusual for a large building to have more than one type of bracing system or core, depending upon the structural requirements and relative positions of the cores on plan.

Chapter 4

Types of beams, columns and trusses

The main types of structural members that may be encountered in general building construction are described in the following sections. These members are usually concealed or are generally not of architectural significance, but an understanding of the range of structural options is important.

4.1 Beams

Beams are designed to resist bending moments and shear forces. The shapes of hot-rolled Universal Beams (UBs) listed in Table 1.1(a) are designed to achieve optimum bending properties for the use of steel. The proportions of well-designed beams fall within relatively narrow limits, depending on the form of loading. As a rule of thumb, sections with a span:depth ratio of 15 to 18 may be used in the scheme design of uniformly loaded steel beams, i.e. for a span of 10 m, the steel beam will be approximately 600 mm deep.

Steel beams can also be designed to act compositely[10] with a concrete floor slab by use of welded shear-connectors, a technique that has achieved great success in North America and in the UK. Its advantages have been realised in so-called 'fast-track' construction by using steel decking as a working platform, as permanent formwork, and as a composite slab acting with the in-situ concrete (see Section 4.1.4 on composite beams).

4.1.1 Floor grillages

The layout of floor beams in buildings depends largely on the spacing of the columns. Often columns can be spaced closer together along the edges of the building, in order to support the façade elements. The primary beams span between the columns, and support secondary beams which then support the floor slab. In most buildings with regular bays, the primary beams support more load than the secondary beams, and are therefore heavier and generally deeper. However, in buildings with unequal bays (e.g. 6 m × 7.5 m), it is possible to design the primary beams to span the shorter distance between columns, so that primary and secondary beams can be designed to be of similar depth.

The simplest arrangement of members in floor grillages uses Universal Beam (UB) sections with pinned connections. In cases where headroom is limited, such as in renovation applications, Universal Column (UC) sections may be used as shallow, although heavier, beams.

In many buildings, designing longer spans internally, such as by spanning directly between the façade columns, creates more flexible space planning. In this case, a variety of structural systems may be used, either as long-span primary beams or secondary beams. These long-span systems generally use the principles of composite construction to increase their stiffness and strength, and often provide for integration of services within their depth.[11]

Typical floor-beam layouts are shown in Figure 4.1, depending on the aspect ratio and span of the floor grid between columns. Heavier beams should be connected to the column flanges, but this is not always possible, such as the floor grid in Figure 4.1(d). Special

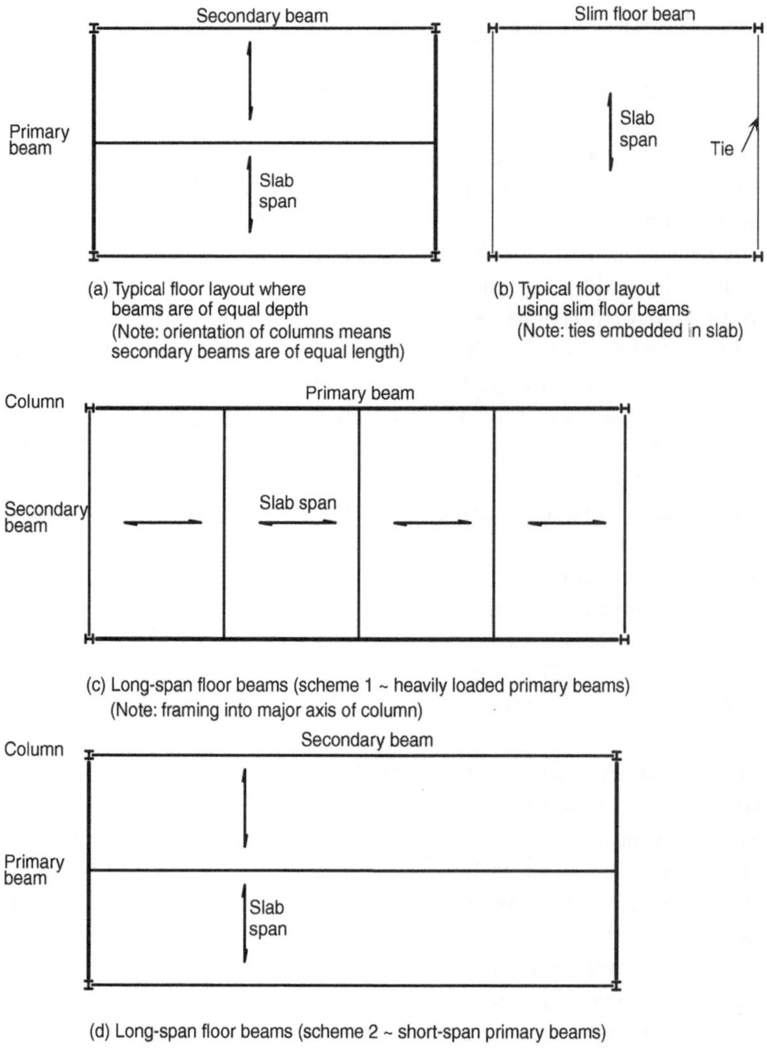

4.1 Typical floor-beam layouts for various spans

detailing measures may be required when wide beams connect to narrower columns (see Sections 5.4 and 5.5).

Slim floor construction, illustrated in Figure 4.1(b), differs from other forms of construction in not requiring secondary beams internally, other than tie members for reasons of stability during construction (see Section 4.1.3). Although slim floor beams are heavier than the equivalent downstand beams, they provide the minimum sensible floor depth, which is broadly equivalent to a reinforced concrete flat slab.

4.1.2 Perforated sections

Castellated or cellular beams are examples of longer span members which have large openings within their depth.[12] These beams achieve the benefits of greater structural efficiency by increasing the section depth for a given use of steel, and provide multiple routes for services. Cellular beams have architectural appeal by their apparent lightness and distinctive appearance in long-span roofs and floors, as in Figure 4.2.

In a castellated beam, the web of a rolled section is cut along the length of the beam in a 'wave' form, as shown in Figure 4.3. The two pieces are separated, offset and then welded together to achieve a

4.2 Curved cellular beam used for architectural effect

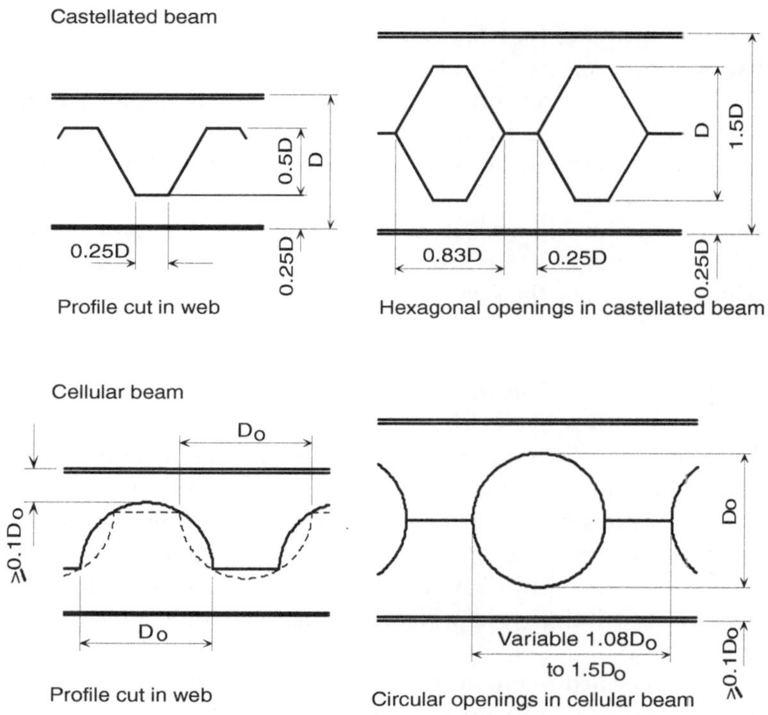

4.3 Profiling of castellated and cellular beams

deeper section. As the weight of steel is unchanged, the structural efficiency of the section in bending is increased. The web, however, is the main source of shear strength and, for this reason, the openings at points of support and/or concentrated load are often filled in using welded plates.

In a cellular beam, also shown in Figure 4.3, the web of a rolled section is cut to form circular or elongated openings. This is an operation in which the profile is shaped in such a way that some of the web is discarded during cutting. Cellular beams are highly efficient and offer many architectural opportunities. The top and bottom parts of the section can be of different sizes, and the sections can be easily adjusted and curved prior to the welding process. The sections can be precambered at no additional cost.

The most appropriate use of castellated or cellular beams is for long spans with moderate loadings, such as in roof structures or in secondary beams in floor grillages. The regular circular openings in a cellular beam are very efficient for distribution of circular ducts in heavily serviced buildings (Figure 4.4). Typical ranges of dimensions are indicated in Figure 4.3. The diameter of the openings can vary between 0.5 to 0.8 times the depth of the beam.

4.4 Cellular beam shows integration of circular service ducts

4.1.3 Slimflor *and* Slimdek *construction*

A slim floor beam is a special case of a modified section where a flat steel-plate is welded to a standard UC section. This generic system is trademarked as '*Slimflor*' by Corus. The plate supports the floor slab so that the beam is partially encased within the floor depth, resulting in a structural system with no downstand beams, leading to reduced floor to floor heights.

Two variations of *Slimflor* construction exist:

- precast concrete slabs spanning between the beams[13] with or without a concrete topping
- deep decking spanning between the beams with in-situ concrete to create a monolithic composite floor,[14] see Figure 4.5.

A series of two-dimensional frames is erected with light steel tie-members between the frames. Spans of the order of 6 to 9 m can be achieved using both variants of *Slimflor* construction.

The decking is designated as SD225, and is 225 mm deep with ribs at 600 mm spacing. This decking is a modern variant of CF210 decking, which is 210 mm deep, and is only used in shorter span applications. The overall floor depth is typically 290 to 350 mm, depending on requirements to control floor vibrations, fire resistance and acoustic insulation.

Slimflor construction was used at New Square, Bedfont Lakes, and the slenderness of its section was expressed elevationally (see Colour Plate 12).

Slimdek[15] offers further advantages in terms of economy and service integration. It consists of a range of rolled Asymmetric *Slimflor* Beams (ASB) and SD225 deep decking which sits on the wider bottom flange (see Figure 4.6). Three ASB sections were produced initially, 280 ASB 100, 280 ASB 136 and 300 ASB 153, which are designated by their approximate height (in mm) and weight (in kg/m). These sections have been designed efficiently for floor grids of 6 × 6 m, 7.5 × 6 m, and 7.5 × 7.5 m respectively, and do not require additional fire protection for up to 60 minutes' fire resistance. A range of 10 ASB sections is now available, including thinner web-beams, which are designed to be fire protected for over 30 minutes' fire resistance. Details of the use of *Slimdek* are given in the Corus *Slimdek* Manual.[16]

Composite action is enhanced by the embossments rolled on to the top flange of the ASB. Openings for services can be created in the ASB between the ribs of the decking. The maximum size of these openings is 160 mm deep × 320 mm wide to facilitate use of flat oval-ducts for services.

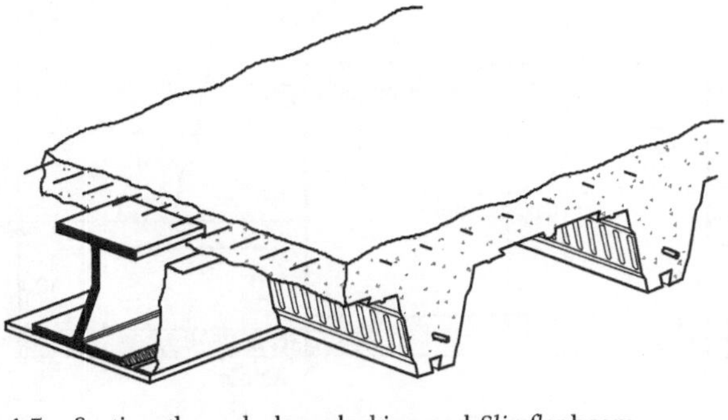

4.5 Section through deep decking and *Slimflor* beam

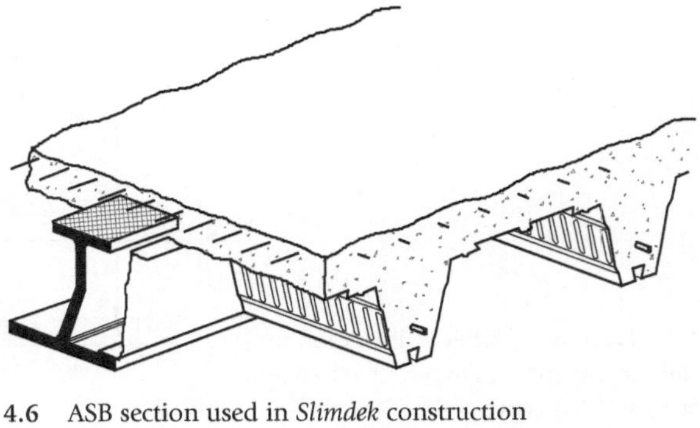

4.6 ASB section used in *Slimdek* construction

The main benefits offered by *Slimdek* construction are:

- reduction in floor depth (by up to 300 mm relative to conventional beam and slab construction)
- no downstand beams, offering ease of service installation
- inherent fire resistance (60 minutes can be achieved without fire protection)
- savings in cladding and services costs
- integration of services and fitments.

The deep decking can be placed rapidly to create a working platform. The space between the ribs of the decking may also be used for ducting, lighting and terminal units. Figure 4.7 illustrates the use of active chilled beams and ducting placed between the ribs which form a continuous line along the building. The overall depth of the floor construction, including a raised floor, is only 600 mm.

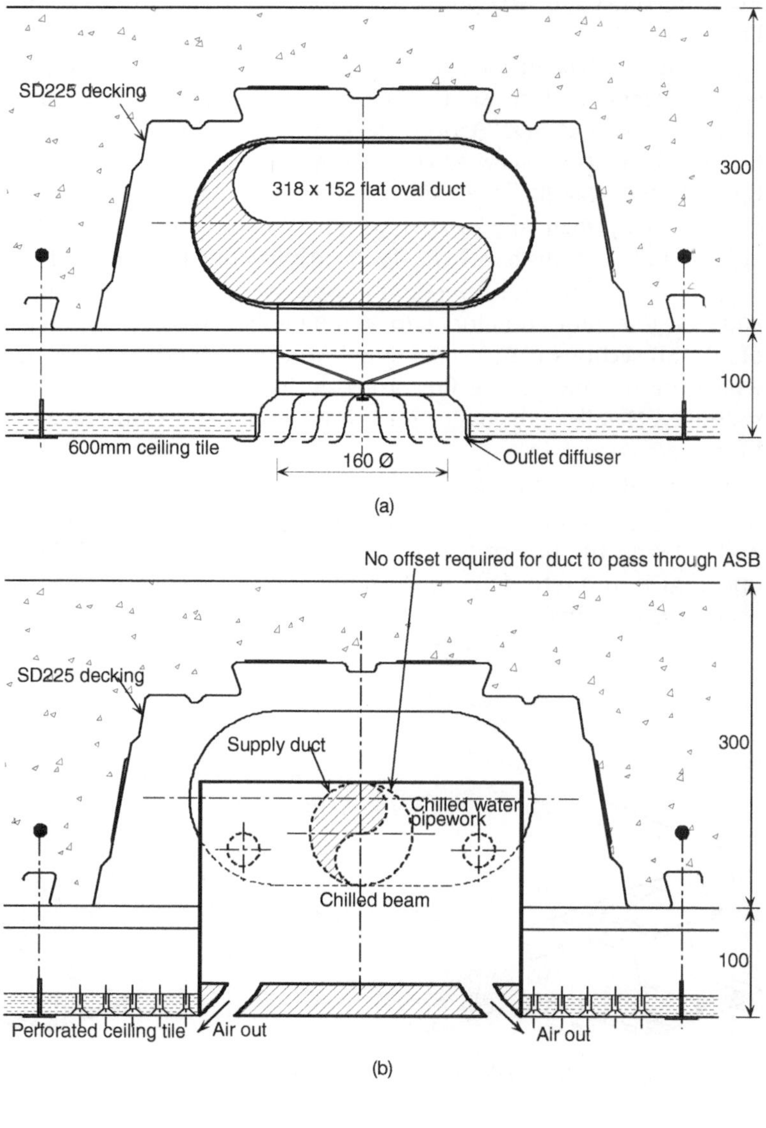

4.7 Example of service integration in *Slimdek*: (a) space between ribs used as a duct; and (b) active chilled beam

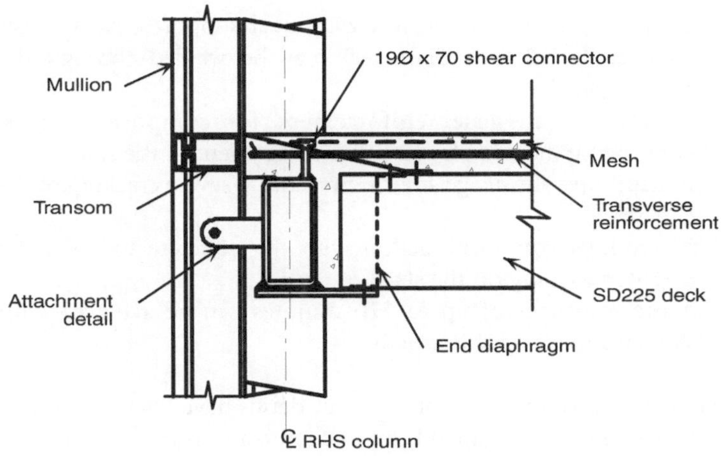

4.8 Use of RHS *Slimflor* beam to support cladding

A rectangular hollow section (RHS) with a welded bottom-plate may be used as a *Slimflor* edge beam.[17] It provides enhanced torsional stiffness to out of balance loads and also presents a 'pencil-thin' edge to the floor which may be visually desirable in some circumstances, such as fully glazed façades. Cladding attachments may be made more easily to the RHS section than to a concrete slab or encased steel-section (see Figure 4.8).

4.1.4 *Composite beams*

Steel beams can be designed to act compositely with a concrete or composite slab by the use of shear connectors, normally in the form of welded studs,[10] attached at regular spacing to the top flange. Composite beams behave essentially like a series of T-beams in which the concrete slab acts as the compression flange and the downstand steel section acts as the tension-resisting element.

Composite action has the effect of greatly increasing the strength and stiffness of a steel beam, and consequently can lead to longer spans for the same size of section or, alternatively, lighter shallower sections may be used for the same load and span configuration. For the efficient design of composite beams, it is often found that the ratio of span to beam depth is in the range 22 to 25, which is approximately 30% shallower than concrete or non-composite alternatives.

Composite decking is usually placed as sheets up to 12 m in length, and is fastened down to all the beams at regular centres. It offers a number of advantages:

- It supports loads during construction without temporary propping up to approximately 3.6 m span.
- Spans of up to 4.5 m can be achieved, if the slab is propped during construction.
- It stabilises the structural members and stiffens the frame against wind loads.
- It provides a safe working platform.
- It acts as a safety net against falling objects.

- To achieve composite action with steel support beams, shear connectors (19 mm diameter) may be welded through the decking on site.
- It acts as transverse reinforcement for composite beams, eliminating the need for heavy reinforcement in the slab.
- It distributes shrinkage strains, preventing severe cracking of the concrete.
- It develops composite action with the concrete to resist the imposed loading on the slab.
- A fire resistance of up to 120 minutes can be achieved with standard mesh reinforcement.

New deck profiles of 80 or 100 mm depth have been developed which extend the span capabilities of composite slabs.

4.1.5 Composite beams with web openings

In composite beams, large openings may be formed through the web. These openings are used for the passage of services within the beam depth, and are about twice the size that would be possible in non-composite beams. The openings are normally rectangular in shape, but may also be circular or square. Welded stiffeners placed horizontally above and below the openings increase the size and aspect ratio of opening that may be used. However, access for welding of the stiffeners may be difficult in shallow beams.

For the scheme design of composite beams with rectangular openings, it is recommended that:[18]

- large openings should be located between one-fifth and one-third of the span from the supports in uniformly loaded beams to have the least impact on the structural design of the beam
- large openings may also be placed close to mid-span, but require over-design of the beam in bending
- the spacing between the edges of openings, or to the connections of secondary beams, should not be less than the beam depth, *D*, unless the effect of these local forces is calculated
- rectangular openings should be not located at less than 2*D* from the support, in order to avoid the effects of high shear and partial shear connection close to the supports
- the suggested maximum depth × length of rectangular openings are:
  $0.6D \times 1.2D$ for unstiffened openings
  $0.7D \times 1.75D$ for horizontally stiffened openings
- horizontal stiffeners should be extended past the opening to provide local-bending resistance at the corners of the opening
- circular openings are more structurally efficient than rectangular openings and may be placed closer together (as in cellular beams).

These detailing rules are illustrated in Figure 4.9, which also shows alternative stiffening arrangements. The use of larger openings can be justified by more detailed calculations. If the beam is over-designed in its bending resistance, large openings can be formed where shear forces are low. For example, deep openings can be provided at mid-span of

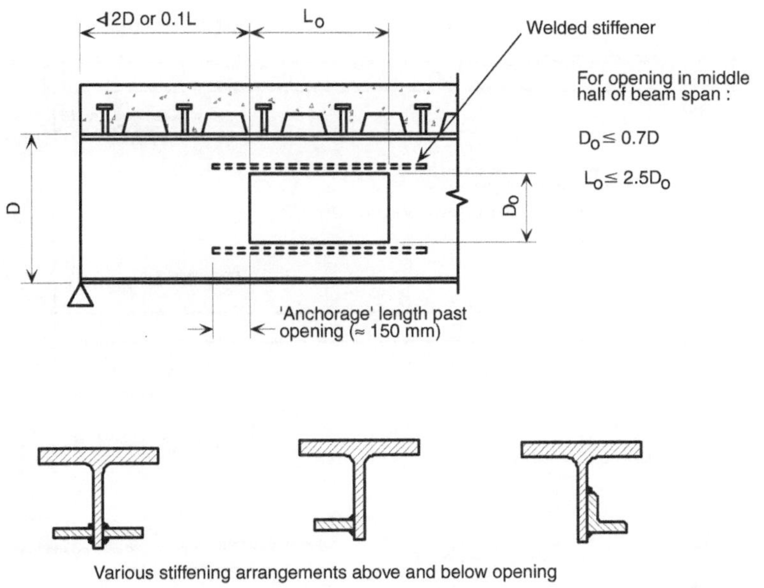

4.9 Detailing of stiffened openings in composite beams

primary beams. Elsewhere, small openings (up to 0.3*D*) can usually be detailed without further checks. Composite beams can also be designed with regular circular openings, as described in Section 4.1.2.

4.2 Long-span beams

Various long-span composite beam systems have been developed in recent years to offer greater provision for services integration. These systems are put into context by presenting the sensible range of spans that may be designed (see Table 4.1). More detailed guidance on structure-service integration is given in a recent SCI publication.[11]

The detailing of each of these structural systems depends on the span and loading configuration. These beams may be designed as either long-span primary beams which are loaded by short-span secondary beams, or more often used as long-span secondary beams which are supported by shorter span primary beams (see Figure 4.1). Primary beams should ideally frame into the major axis of the columns.

Examples of typical designs of the following long-span systems are presented in Figure 4.10. The use of cellular beams and I-beams with isolated web openings were reviewed earlier, and are illustrated in Figure 4.10(a) and (b) respectively.

4.2.1 Stub girders

Stub girders[19] were developed in North America, particularly to meet the needs of deep plan offices with highly serviced space, square grids, and a column spacing of 12 to 15 m. Stub girders comprise a steel bottom chord (normally a UC section) with short steel-sections (stubs) connecting it to the concrete slab. The secondary beams pass over the bottom chord. The openings for services are created adjacent

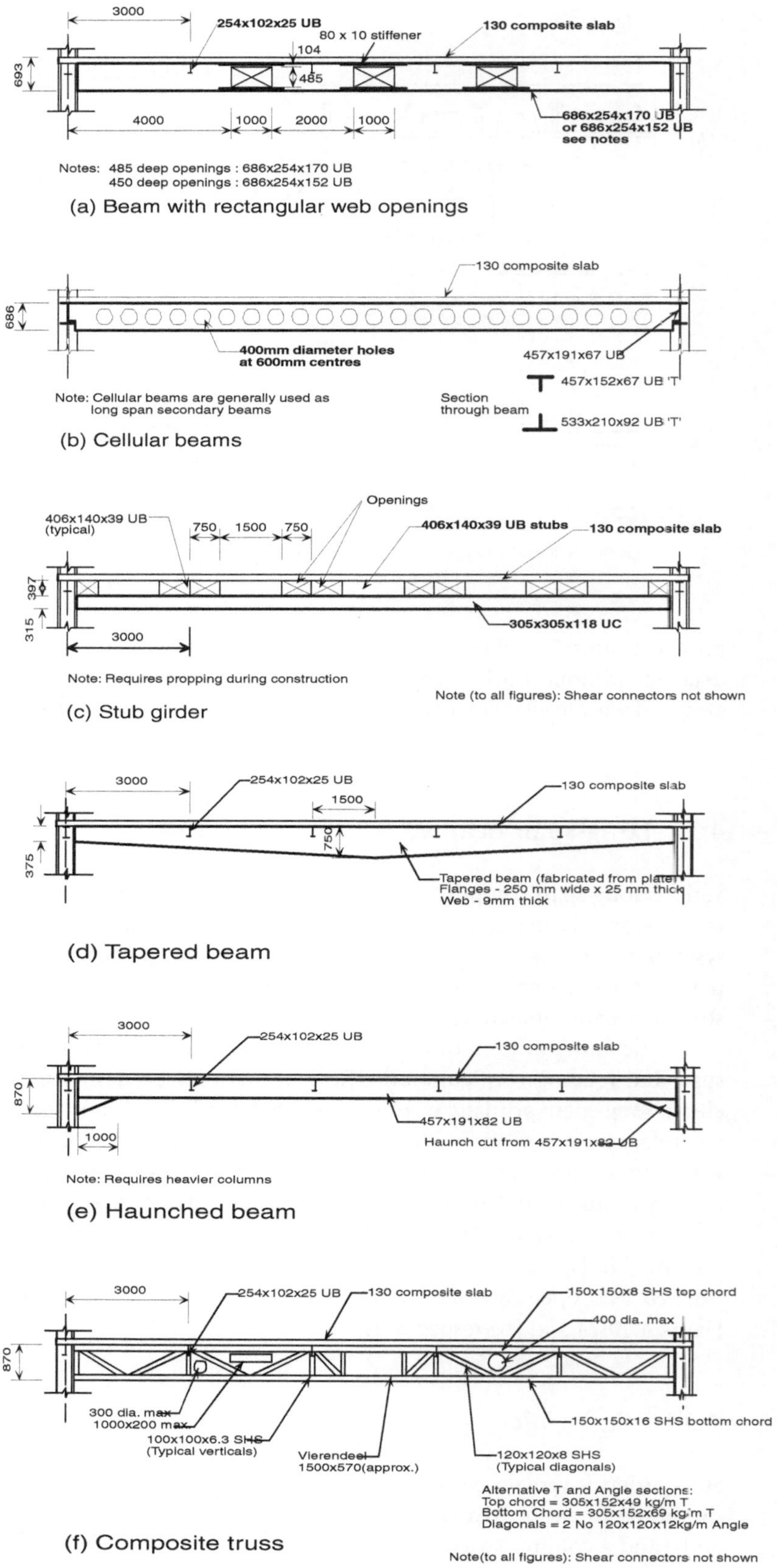

4.10 Long-span composite beams offering the facility for service integration (designed for 15 m × 6 m floor grid)

Table 4.1 Summary of typical spans of different structural systems

Span (m)	6	8	10	13	16	20
Reinforced concrete (RC) flat slab	▬					
Slimdek with deep composite slab	▬	▬				
RC waffle slab	▬	▬				
Shelf angle beam with precast slabs	▬	▬				
Slimflor with precast slabs	▬	▬				
RC beam and slab		▬	▬			
Post-tensioned concrete flat slab		▬	▬			
Composite beam and slab	▬	▬	▬			
Parallel beam approach		▬	▬	▬		
Composite beam with web openings			▬	▬		
Cellular composite beam			▬	▬		
Tapered girder			▬	▬		
Stub girder			▬	▬	▬	
Haunched composite beam			▬	▬	▬	
Composite truss				▬	▬	▬

to the stubs, as shown in Figure 4.10(c). Temporary propping of the bottom chord is often required, although systems with an embedded top chord do not require propping during construction.

4.2.2 *Fabricated sections and tapered beams*

Fabricated girders are made by welding plates together to form I-sections of uniform or asymmetrical cross-section. They are typically used for long-span primary beams or in transfer structures supporting a number of storeys over an open-plan ground-floor area. Webs can be tapered, i.e. of linearly varying depth, so as to create zones for services near to the supports.

A variety of opening shapes may be created in fabricated sections to meet the servicing and architectural requirements. A good example is shown in Figure 4.11. These sections are fire protected by an intumescent coating that was applied off-site to further speed up the construction process.

Tapered beams are also used in long-span portal frames. As illustrated in Figure 4.10(d), tapered beams[20] are particularly suitable for long spans where service zones are created adjacent to columns. Notch tapered or double tapered beams can be designed to optimise these service zones. Large openings may also be provided close to mid-span, where shear forces are low.

4.11 Fabricated beams with multiple openings

4.2.3 *Haunched composite beams*

Frames with rigid connections often use some form of haunched connection,[21] as shown in Figure 4.10(e). The haunch is made from a cut piece of the main beam section, and is designed to achieve a sensible moment connection to the column.

In highly serviced buildings, the main ducts may therefore pass beneath the beams, which are of the minimum practical depth. Typically, the span:depth ratio of haunched composite beams is of the order of 30 to 35 (in terms of the depth of the steel beam). The disadvantage of haunched construction is the need for heavier columns to resist the moments transferred to them. Connections are made to the major axis of the columns in all cases.

4.2.4 *Composite trusses*

Lattice girders or trusses[22] are frequently used in multi-storey buildings and are the most appropriate for spans in excess of 15 m, where the truss is designed to occupy the full depth of the floor and service zone. A high degree of service integration is provided, as shown in Figure 4.10(f). The fabrication and fire protection costs are relatively high, but composite trusses can be cost effective. Some diagonal members may be replaced by vertical members at mid-span so creating a Vierendeel girder to permit the passage of larger services. Trusses can be designed using either CHS or RHS sections in long-span applications. Alternatively, Tee sections may be used as chords, and single or double angles for the diagonal members in medium-span applications.

4.2.5 *Parallel beam approach (PBA)*

In the parallel beam approach (PBA), secondary or 'rib' beams pass over primary or 'spine' beams to form a grillage of members. The primary beams are placed in pairs so that they pass either side of the columns and are attached by large steel-brackets which transfer the shear forces into the column (see Figure 4.12). Secondary beams are

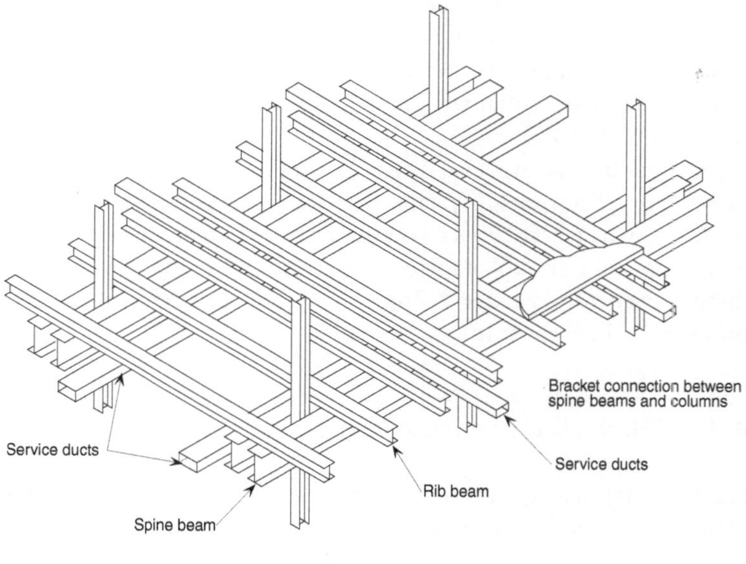

4.12 Parallel beam approach showing service zones

designed to span the greater distance because they can develop composite action with the slab.

The PBA system[23] enables continuity of the beams to be achieved without the high cost of moment-resisting connections. Beam lengths are only limited by handling and transportation requirements. This improves efficiency for long-span applications and can save erection time and costs as the piece count is significantly lower than for conventional construction. Economic comparisons have shown that the PBA system can be advantageous for certain floor layouts for highly serviced buildings. Whilst it may appear that the PBA system would lead to deeper floor construction, their depth remains shallow because of the continuity of the beams, and the overall depths are comparable with the other forms of construction noted earlier.

4.3 Curved beams

Steel members of any standard section can be supplied curved to a constant radius. This is a specialised technique, and advice should be sought on the appropriate bending radii relative to the section which is proposed to be curved. The minimum radius to which any section can be curved depends on the metallurgical properties of the steel and particularly its ductility, cross-sectional geometry and its end use.

Examples of curved beams used in roofs and canopies are illustrated in Figures 4.13 and 4.14.

4.13 Curved channel sections used to create a canopy at Helsinki Airport

4.14 Curved cellular beams in sports facility, Rangers Football Club

4.3.1 *Radii of curving*

Table 4.2 gives the typical radii to which a range of sections can be curved about their major (*x–x*) axis. These are considered as minimum radii, although tighter bends may be considered for some uses. Within one serial size, heavier weights (or thicker walls) can be curved to smaller radii than lighter ones. Sections can be curved about the major or minor axis, and reversed curves and hoop shapes are also possible.

The minimum radius to which sections can typically be bent varies from less than 1 m for the smaller sections to some 50 m for the largest. The normal tolerance for a particular radius of curve is 8 mm. Curving is performed by 'cold' bending, which involves passing the member through a set of three rollers. These rollers are purpose-made to match the precise shape of the cross-section being bent in order to avoid local buckling of the section. Larger sections can be heated locally (induction heating) to achieve smoother, tighter curves.

A general fabricator may be able to perform such work, but most curving is performed by specialists who have developed their own range of bending rolls to enable the entire range of sections to be curved.

4.15 Cellular beams used in an Asda store, Tamworth

4.3.2 *Curved beams in roof structures*

Curved roof structures provide many architectural opportunities for expression, particularly where the walls and roof are combined in one overall structural solution, so that the demarcation between these elements is removed. The physical nature of these roofs or enclosures is that they are curved to a radius to allow water run-off and to achieve maximum usable space internally.

Castellated and cellular beams have been used successfully in long-span roofs when curved and with intermediate supports (see Figure 4.15). Although slightly curved in shape, these members function as beams. Here the lightness of the highly perforated sections is combined with the ability to curve the sections in the re-welding process. The rhythm of the regular openings in the beams provides a degree of architectural interest.

Curved sections may also be used to great effect to create arch structures, as in the Lea Valley Ice Skating Rink in Figure 4.16, which was one of the first structures of this type. In this building, the deep-profiled decking was designed as a 'stressed skin' to eliminate the need for secondary bracing members, and to accentuate the simplicity of the structure.

4.16 Curved roof at Lea Valley Ice Skating Rink (architect: BDP)

4.3.3 *Curved tubular members*

Curved roofs may be formed using single-curved tubular members, double-layer members connected periodically, or tubular trusses, as illustrated in the following examples.

Table 4.2 Minimum bending radii for common steel sections

	Section	Typical radius
	**Joists and Universal Beams (x–x axis)**	
	610 × 305 × 238 kg/m UB	40.0 m
	533 × 210 × 122 kg/m UB	30.0 m
	406 × 178 × 74 kg/m UB	15.0 m
	254 × 203 × 82 kg/m RSJ	3.5 m
	203 × 152 × 52 kg/m RSJ	2.2 m
	152 × 127 × 37 kg/m RSJ	1.6 m
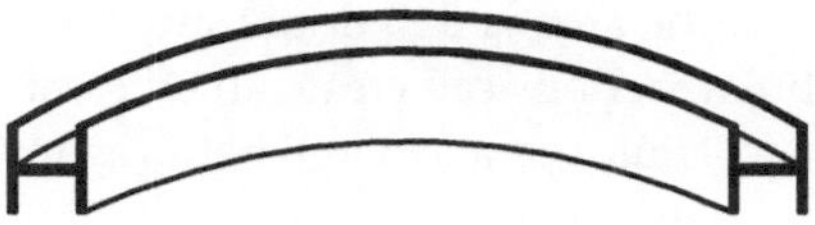	**Universal Columns (x–x axis)**	
	152 × 152 × 37 kg/m	2.0 m
	203 × 203 × 86 kg/m	3.0 m
	254 × 254 × 167 kg/m	4.0 m
	305 × 305 × 283 kg/m	5.0 m
	All sections up to 356 × 406 UC	
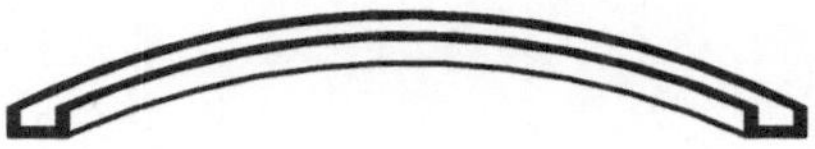	**Channels (x–x axis)**	
	127 × 64 × 14 kg/m	2.0 m
	203 × 89 × 29 kg/m	3.0 m
	254 × 89 × 35 kg/m	5.0 m
	305 × 102 × 46 kg/m	7.0 m
	All sections up to 432 × 102 × 65 kg/m	
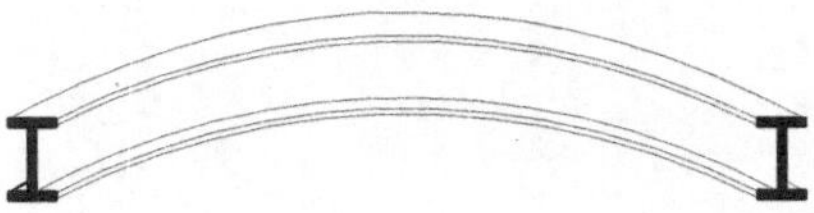	**Joists, beams and columns (y–y axis)**	
	127 × 76 × 16 kg/m	0.8 m
	230 × 133 × 30 kg/m	1.5 m
	457 × 191 × 98 kg/m	2.5 m
	610 × 229 × 140 kg/m	3.0 m
	All sections up to 1016 × 455 × 488 kg/m	
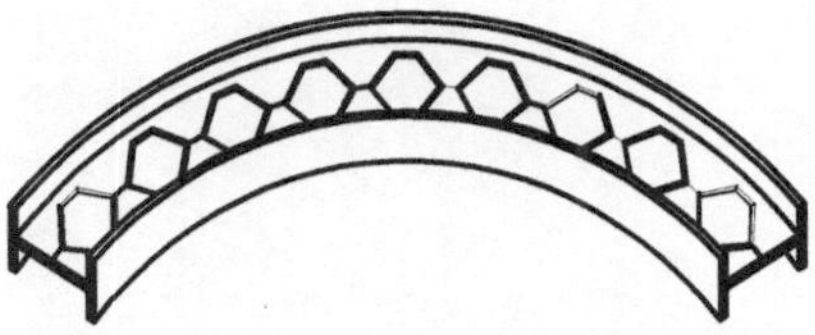	**Castellated and cellular beams (x–x axis)**	
	305 × 133 × 30 kg/m	4.0 m
	458 × 165 × 54 kg/m	8.0 m
	609 × 178 × 74 kg/m	12.0 m
	800 × 210 × 122 kg/m	16.0 m
	915 × 305 × 238 kg/m	20.0 m
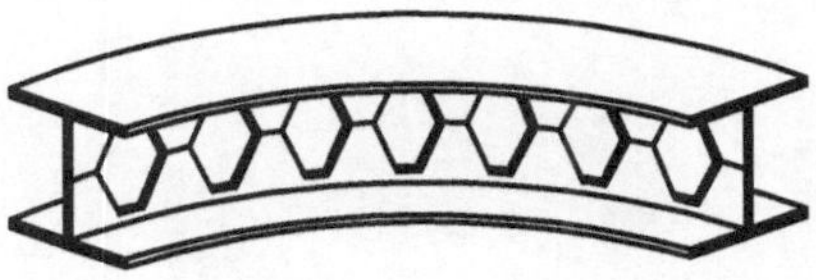	**Castellated and cellular beams (y–y axis)**	
	305 × 133 × 30 kg/m	2.0 m
	458 × 165 × 54 kg/m	3.0 m
	609 × 178 × 74 kg/m	4.0 m
	800 × 210 × 122 kg/m	6.0 m
	915 × 305 × 238 kg/m	8.0 m
	**Circular Hollow Sections**	
	60.3 × 5 mm	0.4 m
	114.3 × 6.3 mm	0.7 m
	168.3 × 10 mm	0.9 m
	219.1 × 12.5 mm	1.1 m
	Most sizes up to 610 mm o/d × 35 mm	
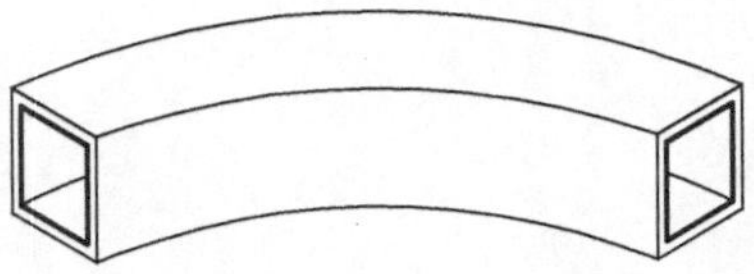	**Square and Rectangular Hollow Sections**	
	50 × 50 × 5 mm	0.6 m
	100 × 100 × 6.3 mm	1.1 m
	150 × 150 × 10 mm	1.4 m
	200 × 200 × 12.5 mm	2.0 m
	All sizes up to 400 × 400 × 16 SHS and 500 × 300 × 20 RHS	

The Law Faculty at Cambridge is enclosed by a triangulated Vierendeel structure, which is cylindrical in section, to which the glazing and stainless steel cladding is fixed. This curved element merges between the roof and walls, and provides a sympathetic enclosure to the internal space, as illustrated in Figure 4.17 and see also Colour Plate 1.

The TGV station at Lille extends the concept of a curved tubular roof by using ties periodically to decrease the bending effect in the arch, so as to minimise the required size of the sections (see Figure 4.18). For the Amenity Building of the Saga Headquarters, a series of inclined arches support the fabric roof, as illustrated in Figure 4.19.

A three-dimensional extension of the arch is to a dome structure, where a latticework of welded tubular sections can create an efficient and attractive structural solution to auditoria and large halls, as in Figure 4.20.

The roof of the great glasshouse of the National Botanical Garden of Wales (Colour Plate 2) comprises a series of tubular steel arches in the form of a toroid. The attachment to the glazing is made by projecting fins welded to the tubes.

4.17 Curved enclosure at Cambridge University Law Faculty (architect: Foster and Partners)

4.18 TGV station at Lille, showing the use of a tubular tied arch structure

4.19 Amenity Building, Saga HQ (architect: Michael Hopkins and Partners)

4.20 Roy Thomson Hall, Toronto, use of welded tubular lattice to create the dome-like roof of the concert hall

4.21 Leipzig Messe — external structure supporting glazing (architect: Von Gerkan Marg & Partners and Ian Ritchie)

At the Leipzig Messe, the curved tubular arches support the tubular latticework which supports the fully glazed façade, as shown in Figure 4.21 and see also Colour Plate 6.

4.4 Columns

In braced frames, columns are designed to resist mainly compression forces. The shape of UC sections is such that they are more efficient in resisting buckling than standard beam sections. Columns used in rigid or sway frames are also designed to resist bending. Where bending effects are dominant, it may be more appropriate to use UBs as columns, such as in portal frames.

- *UC sections*
  Columns may be in the form of UC sections that are spliced at appropriate points (usually every two or three storeys) in tall buildings. In taller buildings, column sizes are generally selected from one serial size with decreasing section weight at upper levels. Beam to column connections are made either to the flanges of the column (major axis connections) or to the web of the column (minor axis connections). Illustrations of typical connections are given later. It may also be necessary to stiffen the columns locally at points of load transfer, such as for beams with moment connections.

- *Tubular columns*
  Square or Circular Hollow Sections are very efficient in compression because the material is remote from the axis of the section, therefore increasing the resistance to buckling. Both circular (CHS) sections and square (SHS) sections are widely used as slender columns. The main design issue is the form of connection to the face of the column.

- *Composite columns*
  Columns may be designed to achieve greater compression and fire resistance by concrete encasement (in the case of I-sections) and concrete filling (in the case of hollow sections). For example, the in-filling between the flanges of an I-section column without reinforcement can increase its fire resistance up to 60 minutes, whilst retaining the same external dimensions of the section.[24] The in-filling of tubular sections with concrete can increase their structural resistance[25] and also their fire resistance to up to 60 minutes without reinforcement, and up to 120 minutes with bar reinforcement.

4.4.1 *Exposed tubular columns*

Tubular columns are used in applications where the minimum amount of intrusion into the space is sought, or where the external appearance of the column is preserved. They are structurally efficient and can thus be used to advantage in slender columns. Amsterdam's

Schiphol Airport illustrates this principle, as shown in Figure 4.22. The large diameter columns were also used as part of the air-ducting system. Hong Kong's Hung Hom station shows how tubular columns can be used within the tubular spine beams (see Figure 4.23). For another example of exposed tubular columns see Colour Plate 7.

A recent example of clustering columns together in an attractive manner can be seen in the Mediatheque Centre at Sendai (Colour Plate 21).

Circular columns are particularly attractive internally in shopping malls and auditoria. Fire resistance can be achieved by the use of intumescent paints or by concrete filling (see Chapter 13). Other forms of fire protection normally affect the appearance and shape of the section and are not preferred.

Connections to tubular columns are often expressed as part of the overall structure concept. As described in Chapter 6, there is a wide

4.22 Amsterdam's Schiphol Airport

4.23 Tubular columns and spine beams in Hong Kong's Hung Hom Station (architect: Foster and Partners)

range of structural options, depending on whether the connection is pinned (i.e. resists only shear and tension) or rigid (i.e. also resists moment). Examples of architectural details used in beam to tubular column connections are illustrated in Figure 4.24.

Columns are erected in two or three storey high lengths and splices are usually made by end plates or similar connections just above floor height, in order to avoid intrusion into the floor space.

Tubular sections may also be used as heavily loaded struts to support canopy roofs, as shown in Figure 4.25. This is important also in wind uplift conditions where the reversal of loading may cause tension members to act in compression.

Atria and shopping malls often use tall and slender tubular columns to support long-span roofs, as shown in Figure 4.26.

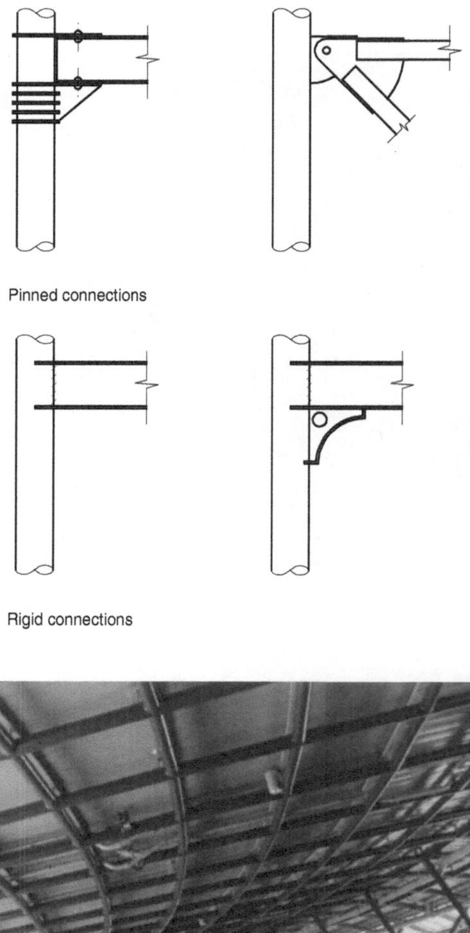

4.24 Tubular columns with pinned or rigid connections

4.25 Tubular struts used to support the roof of Wimbledon No. 1 Court (architect: BDP)

4.26 Curved roof at Princes Square, Glasgow (architect: Hugh Martin & Partners)

4.4.2 *Concrete-filled columns*

Concrete filling improves the compressive resistance and fire resistance of tubular columns. This is because the concrete within the section acts compositely with the steel casing, so that the compressive strengths of the two materials can be mobilised together. Indeed, the strength of the concrete is also enhanced by the confining effect of the tubular section. Often the designer does not utilise the compressive strength of the concrete in normal design, but uses it to enhance the fire resistance of the column on the assumption that the exposed steel section loses all its strength in a severe fire.

A design method for composite columns is presented in an SCI publication.[25] From the point of view of architectural opportunities, concrete filling can lead to:

- more slender columns
- more heavily loaded columns, where the compressive resistance is increased for a given tube size
- longer fire resistance periods (which can also be enhanced by bar reinforcement)
- excellent impact resistance.

In Australia and the Far East, large diameter concrete-filled tubular sections have been widely used in high-rise commercial developments. In this case, the tubular sections are designed principally to support the framework and floors during construction, and the concrete provides the compressive resistance to subsequent loads. Large tubular sections (of 0.6 to 1.5 m diameter) can be produced from plate, which is bent into a circular form and welded along its seam.

Particular technical issues to be addressed in this form of construction are:

- the method of concrete filling, which is normally by pouring from the top of the column in one or two storey heights
- load transfer from the beams to the columns, which for large diameter columns is achieved by a steel insert with shear connectors embedded into the core of the column
- fire resistance by bar reinforcement in the concrete. In this case, the column is designed and detailed to standard reinforced-concrete practice. The amount of reinforcement should be minimised (<2% cross-sectional area of the column) in order not to be too congested for concrete filling.

4.4.3 *Tubular masts*

Tubular members can be used in tall slender masts and can be combined with other sections, depending upon the architectural and structural approach. One excellent early example of the combined use of section types is the Renault Parts Distribution Centre in Swindon, where circular tubular columns supported a framework of tapered UB sections suspended from the column apex and shaft (see Figure 1.2).

Single or clustered tubular columns may themselves form a basic structure with opportunity for architectural expression. Figure 4.27 illustrates a group of tapering tubes arranged to emulate a ship's crane in Genoa, Italy, which support both a fabric membrane roof over a public piazza as well as an elevator ride to provide panoramic views of the city.

A supermarket canopy in Plymouth (Figure 4.28) was designed using tubular columns to express a nautical theme.

Sometimes the separation between column and truss or beam elements is less clear, as in the case where 'tree-like' structures are devised. This kind of expression was also used at Stuttgart Airport (Figure 1.12) and in a more formal manner at Stansted Airport (Figure 4.29), in which 36 column trees act as a rigid framework and support inclined branches which themselves support the entire roof.

4.27 Cigar-shaped columns used in the Columbus International Exhibition Centre, Genoa (architect: Renzo Piano)

4.28 Sainsbury's supermarket canopy, Plymouth — expressing a nautical theme (architect: Dixon & Jones)

4.29 Column trees at Stansted Airport (Architect: Foster and Partners)

4.5 Trusses and lattice girders

Trusses and lattice girders can be conceived of as triangular or rectangular assemblies of tension and compression elements. The top and bottom chords provide the compression and tension resistance to overall bending, and the web or bracing elements resist the shear forces. A wide variety of forms of trusses can be created. Each can vary in overall geometry and in the choice of the individual elements which comprise them.

Trusses are generally associated with pitched roofs and are designed to follow the roof profile. Shallower roof pitches result in heavier compression chords, whereas steeper roof pitches involve longer and often heavier bracing members.

Lattice girders are generally associated with long-span beams in which the top and bottom chords are usually horizontal. However, for flatter roof pitches, lattice girders with a sloping top chord can also be used efficiently.

4.5.1 Forms of trusses

Trusses or lattice girders may take a number of basic forms, as illustrated in Figure 4.30. The common names for these truss forms are given, together with their typical span range. They are fabricated by bolting or welding standard sections together. For spans of up to 20 m, it is sufficient to use angles, tees and lighter hollow sections. For very long spans, UC or heavier hollow sections may be required. The mixed use of these sections may be appropriate to minimise the visual impact of the bracing members. These alternative section types are shown in Figure 4.31. Trusses are very efficient in the use of steel, but are relatively expensive to fabricate. The bracing members are usually lighter than the chord members.

- *Warren or Pratt Lattice Girders*

  Lattice girders have broadly parallel top and bottom chords in which the bracing (diagonal) members are arranged in a W or N form, respectively. In a Pratt girder (N form), the orientation of the bracing members normally changes at mid-span. The top chord is generally designed to be restrained against out-of-plane buckling by the regular attachment of roof purlins or of the floor slab.

  A striking example of a structure formed from what is essentially a circular three-dimensional Warren girder is the London Eye, designed by Marks Barfield Architects (Colour Plate 23). See also Colour Plate 4.

  Pratt girders are a traditional form of construction often using angle and T-sections. They are efficient at supporting vertical loads because all the compression members are short (i.e. the vertical members) and the longer diagonal members are in tension.

  Warren girders (W form) are often fabricated from tubular sections as they are efficient as bracing members which act

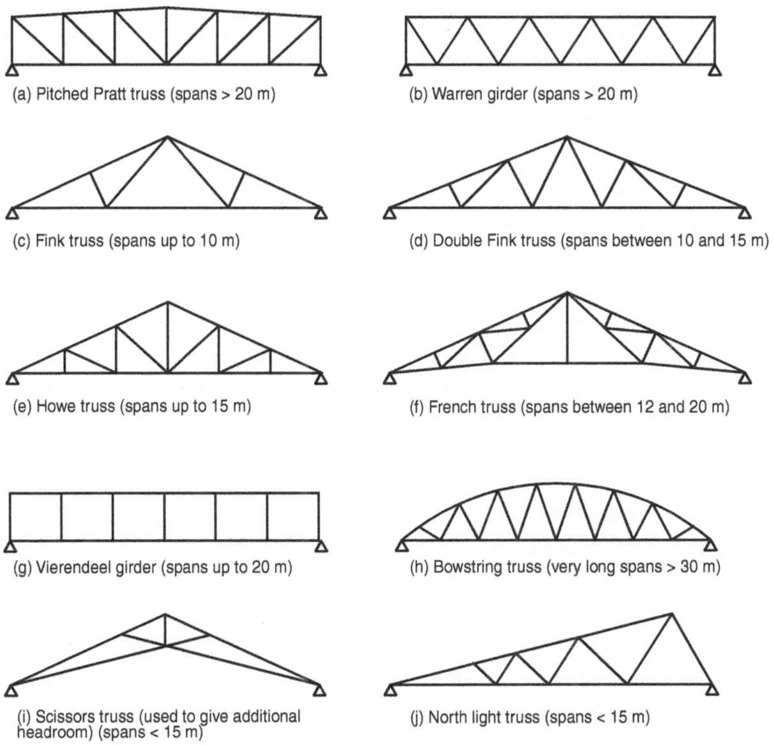

4.30 Different forms of conventional roof trusses and lattice girders

alternately in tension and compression. In lightweight buildings, wind uplift can be significant and may cause reversal of the forces acting on the truss.

- *Fink, Howe and French trusses*
  These particular shapes of pitched truss form the shape of the finished roof. The apex and eaves joints between the chords are pinned. They are often used in housing and modest span roof trusses, and generally comprise Tees and angle members.
- *Vierendeel girder*
  This is a different form of structure in which the diagonal bracing members are eliminated, and the connections between the horizontal and vertical members are made moment-

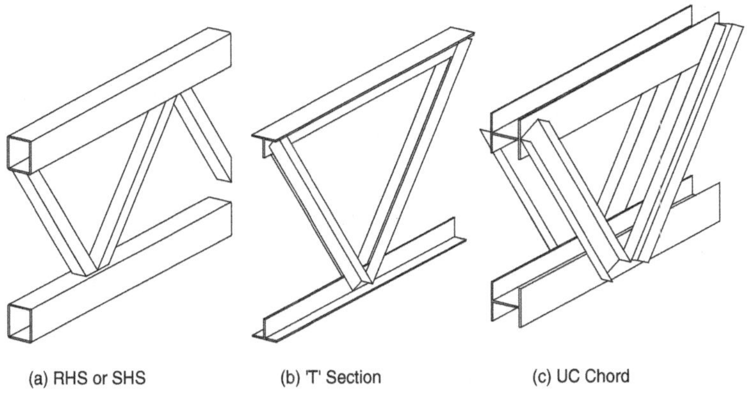

4.31 Different types of steel section used in trusses

resisting. Vierendeel trusses are expensive in the use of steel and in fabrication, and are only appropriate for use in special circumstances, such as when the size of the openings is maximised to permit the passage of services. However, it is possible to design one Vierendeel panel in the centre of an otherwise standard Warren or Pratt girder, especially if the girder achieves composite action with a floor slab.

- *Bowstring truss*
  One chord of a bowstring truss is curved in elevation and is tied between its supports. Light trusses of this form may also be orientated vertically to support cladding and glazing where architectural expression of the truss is particularly important.
- *Scissor truss*
  The scissor truss is a variant of a standard truss form and offers architectural possibilities and greater headroom, but is structurally less efficient because of its shallower depth.
- *North light roof truss*
  North light trusses are traditionally used for short spans in industrial workshop-type buildings. They allow maximum benefit to be gained from natural lighting by the use of glazing on the steeper pitch which generally faces north or north-east to reduce the solar gain.
- *Developments of roof form*
  Most of the above lattice girders and trusses can be further developed into more interesting structural and architectural forms. Some possibilities, including curved and mansard roofs, are illustrated in Figure 4.32.

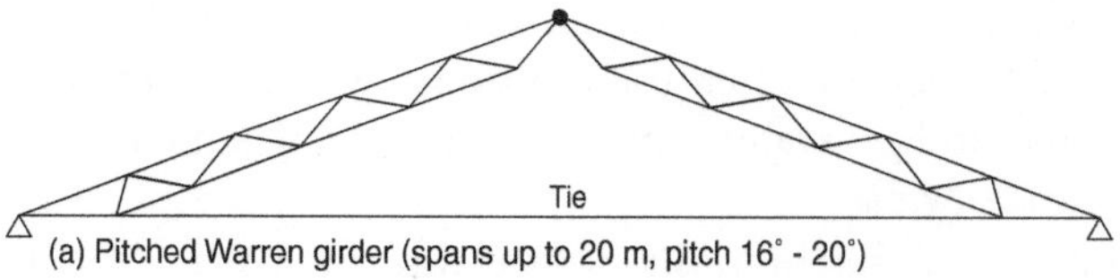

(a) Pitched Warren girder (spans up to 20 m, pitch 16° - 20°)

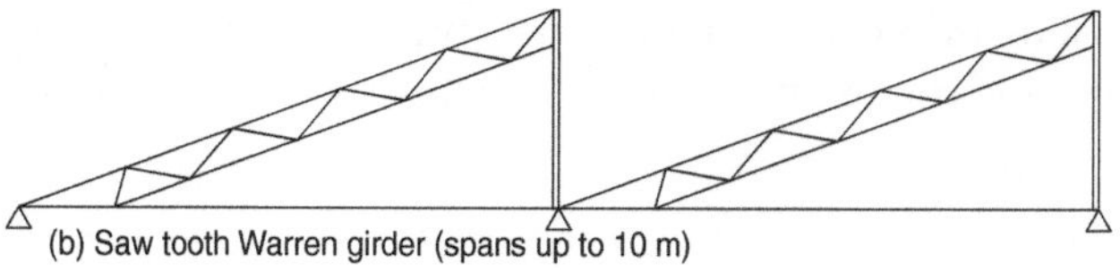
(b) Saw tooth Warren girder (spans up to 10 m)

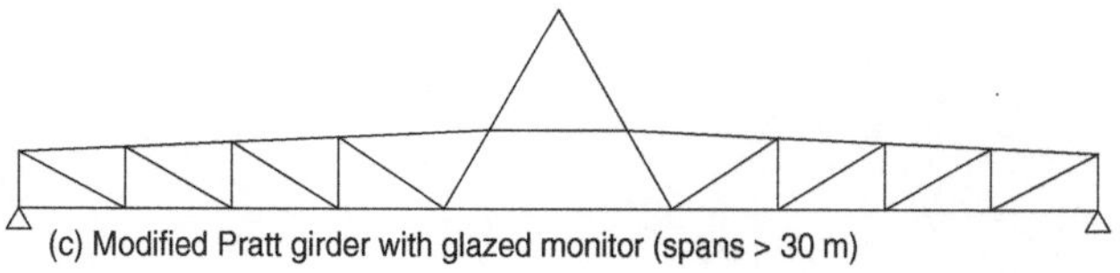
(c) Modified Pratt girder with glazed monitor (spans > 30 m)

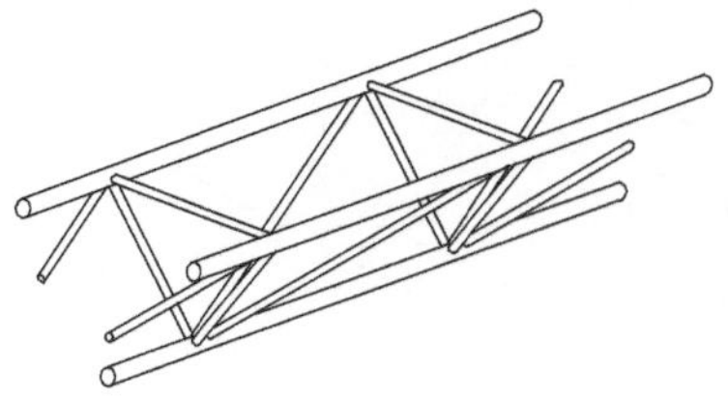
(d) 3-dimensional Pratt girder

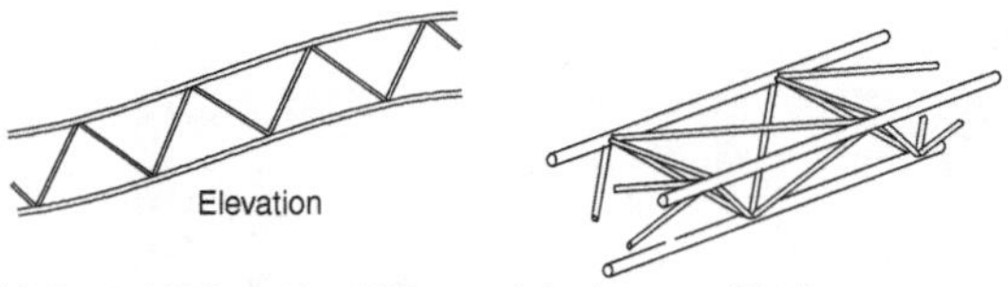

(e) Curved 3-dimensional Warren girder (spans > 20 m)

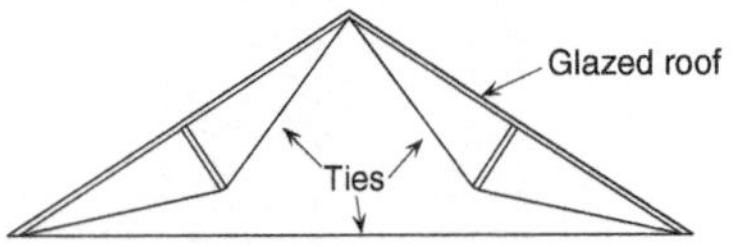

(f) Tied rafter truss (spans < 15 m)

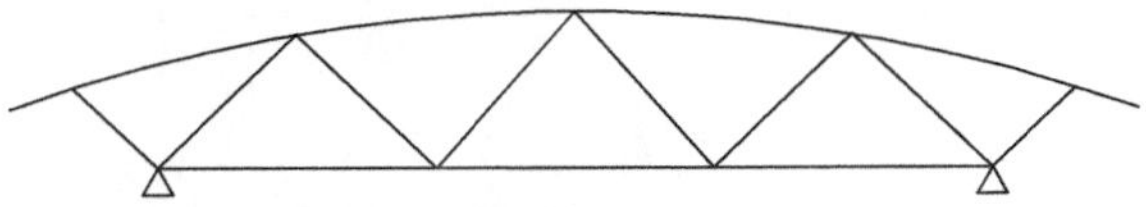

(g) Modified curved Warren girder

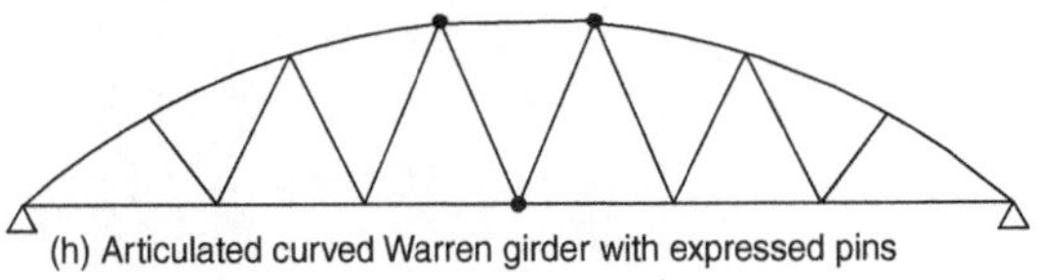

(h) Articulated curved Warren girder with expressed pins

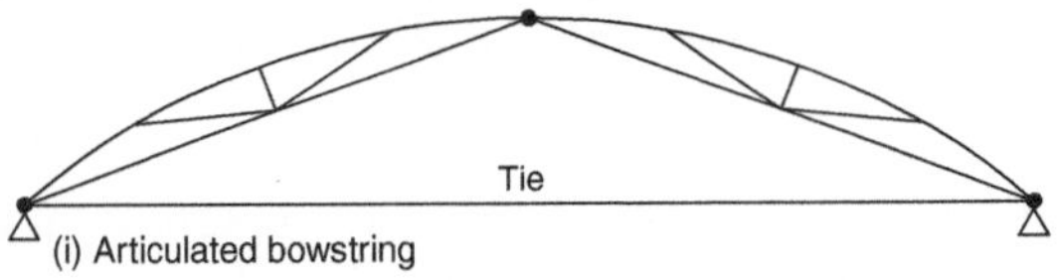

(i) Articulated bowstring

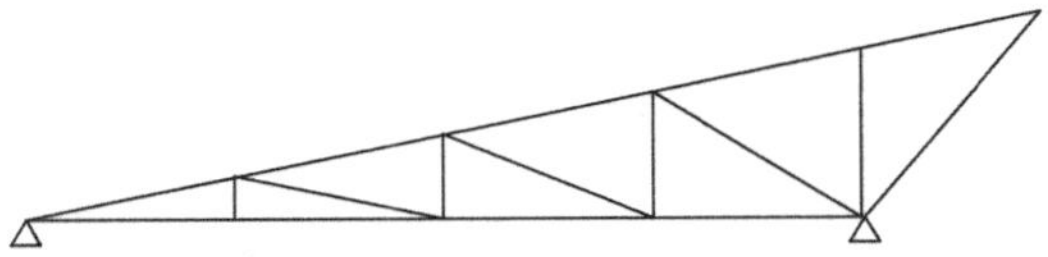

(j) Cantilevered mono-pitch truss

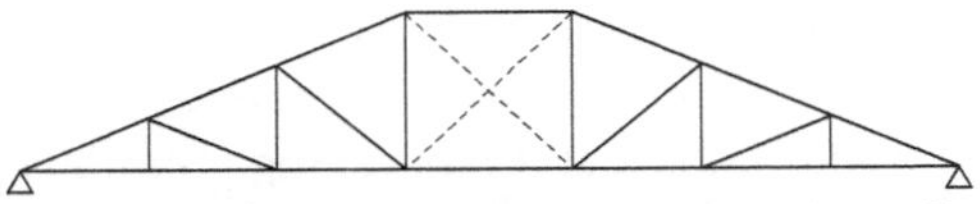

(k) Modified Pratt truss with Vierendeel or braced central bay

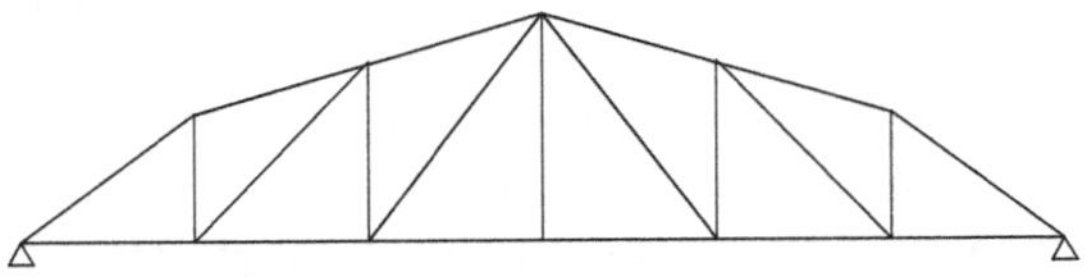

(l) Modified Pratt truss as a mansard

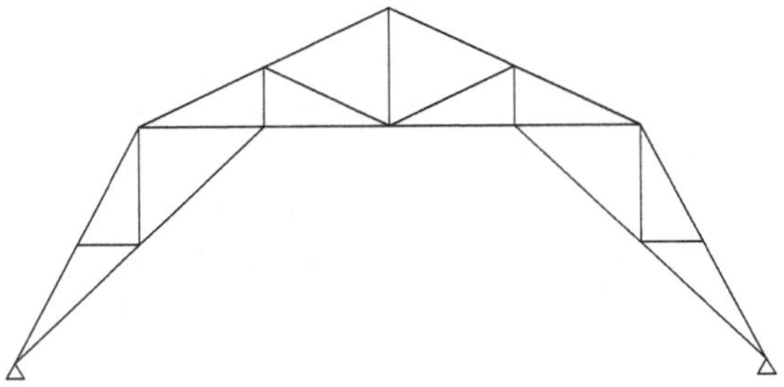

(m) Mansard truss creating open space

4.32 Development of standard truss and lattice forms

4.33 Lattice girders combine with fabricated steel columns making a hybrid-portal structure at the Brit School, Croydon (architect: Cassidy Taggart)

Trusses offer an excellent opportunity for architectural expression in a variety of forms, as illustrated in Figures 4.33 and 4.34.

Other smaller-scale components may be considered depending on the form of the truss, such as:

- *Cables*
  Cables (or steel ropes) are spun from a number of strands or collection of wires. The cables can be impregnated and sheathed with nylon or PVC, and can also be greased and galvanized for corrosion protection. Cables have high-tensile strength but often low ductility. They are suitable only for tensile components in trusses, for example in wind-resisting girders for glazed walls.
  Fitments to the ropes provide the coupling mechanism to the adjacent structure. Special consideration is required to the aerodynamic damping of long ties on cables when exposed to wind.
- *Rods*
  Individual rods are made from solid steel, whose ends are threaded to attach to steel couplers. Rods are linear and more rigid elements, whereas cables will sag naturally. Rods are usually lightly tensioned on erection of the frame. They are only suitable for resisting tension. In a 'wind girder', they can be pretensioned so that the reversal of wind loads does not cause compression.
- *Flats*
  Steel flats may be considered in X-braced trusses, although they are visually more obtrusive.

4.34 Lattice bowstring truss of UK pavilion, Expo 1992, Seville (architect: Nicholas Grimshaw & Partners)

4.5.2 *Articulation of elements within trusses*

The same notions that guide the relationship between members in a frame to give scale, emphasis and articulation to the parts, are equally important to the relationships between elements in an individual member of fixed overall geometry and end conditions. The point is illustrated diagrammatically for a simple planar truss in Figure 4.35 in which the position of the pinned connections between the tension and compression elements, and within the compression elements themselves, can create different details and effects. This principle is generally applicable to any type of member. The particular form assumed by the connections varies depending upon the cross-section of the individual elements.

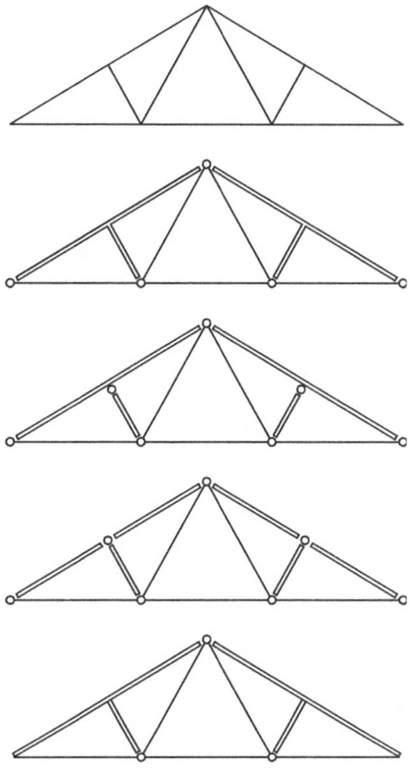

4.35 Articulation of elements within the truss to create different effects

4.5.3 *Tubular trusses*

Trusses using tubular members can provide elegant structural solutions in long-span roofs. They can also be used as 'transfer structures' to support a number of floors above and to create open circulation areas beneath. The span to depth ratio of long-span trusses using tubular sections may be in the range of 20 to 25, reducing to 10 to 15 for heavily loaded applications. Tubular trusses can be very simple in form, as shown in Figure 4.36, which illustrates the use of

4.36 Trusses at Toyota HQ, Swindon (architect: Sheppard Robson)

parallel chord trusses. Inclined tubular trusses may be used in a 'folded plate' form to reflect the shape of the roof, as shown in Figure 4.37. Horizontal forces are resisted by ties (refer to Section 7.5).

More complex roof trusses can be created which are triangular in cross-section, as in Figure 4.38. The bowstring truss in the sports hall shown in Figure 4.39 used a heavy top chord and vertical posts with light bracing and bottom chord members. The apparent depth

4.37 Inclined tubular trusses to create a folded plate roof (architect: Haworth Tompkins Architects)

4.38 Curved triangular trusses at Swindon's Motorola factory (architect: Sheppard Robson)

4.39 Bowstring truss at sports hall

4.40 Stratford Market depot, London (architect: Wilkinson Eyre)

4.41 Deep curved roof trusses at the TGV terminal at Charles de Gaulle Airport, Paris (architect: Aeroports de Paris)

of the bowstring truss is reduced by the use of these lightweight components.

The long-span trusses at Stratford Market depot are arranged in an intersecting orthogonal pattern and are supported on column trees to minimise the effective span of the trusses (see Figure 4.40). At Ponds Forge, Sheffield, the roof trusses were orientated diagonally across the enclosure and supported on a diagonal grid of tubular members (see Colour Plate 20).

The roof trusses to the TGV terminal at Charles de Gaulle Airport, Paris, used tubular trusses comprising a downward curved bottom chord, which is normally the opposite configuration to that desired for most roofs, but which causes a striking architectural effect. The inclined tubular columns support the upper chords of four trusses, as illustrated in Figure 4.41.

4.42 Triangular roof trusses at Hamburg Airport (architect: Von Gerkan Marg and Partners)

At Hamburg Airport the triangular trusses were curved along their length and were supported by inclined tubular struts, as shown in Figure 4.42.

Chapter 5

Connections between I-sections

5.1 Introduction to connections

In a typical steel structure, the detailed design of the connections, the preparation of the production drawings, the fabrication and the erection accounts for some two-thirds of the total cost of the framework. Most of the cost is absorbed in the detailing and fabrication of the connections between the members.

In most projects, steel fabricators undertake the design and detailing of all connections according to their preferred method of fabrication. Because of this, there has tended to be a diversity both in connection types and design methods. Therefore, designers should have an awareness of the range and types of typical details in order to assess the suitability and adequacy of the proposed details.

The provision of industry standards for connection design is developing rapidly. In 1991, the SCI and the British Constructional Steelwork Association (BCSA) first published a design guide, *Joints in simple construction*, which presents design procedures for connections for use in buildings designed with braced frames. A further publication on moment-resisting connections was produced in 1995.[6] The publication on simple connections was revised in 2002[5] and now includes tubular connections.

Various forms of connections may be identified in regular steel frames. These are connections of:

- beams to column flanges
- beams to column webs
- beam to beam webs
- column to column splices
- column bases at foundations
- bracing connections.

Their detailing depends on the forces and moments to be transferred, and on the chosen member sizes. However, some common detailing rules apply, which are discussed in the following sections in order to gain an appreciation of the form of connections in regular frames.

5.2 Benefits of standardisation

In a typical braced multi-storey frame, the elements in the connections may account for less than 3% of the frame weight and yet probably 30% or more of the total cost. Efficient connections will therefore have the lowest detailing, fabrication and erection cost, and are standardised over a broad range of application.

Parameters relevant to standardisation include the:

- type of connection
- grade of steel used in the connecting parts
- bolt grades, sizes and lengths
- weld types
- member sizes and geometry.

The benefits of this approach include:

- a reduction in the number of connection types, making fabrication easier and cheaper
- the development of design aids and associated computer software
- savings in buying, storage and handling time, which leads to a reduction in overhead costs
- reduced design costs, and fewer difficulties in Building Regulation approval
- the use of one grade and diameter of bolt in a limited range of lengths saves time changing drills or punches in the shop, and leads to faster erection and fewer errors on site.

In a particular project, it is important to first define a series of connections of 'common' form, whether pre-existing standard connections or those determined as 'standard' for the member types, sizes and arrangements that are encountered in the building structure. This should also extend to the interfaces with other key components, such as cladding and roofing.

5.3 Industry-standard connections

Standards have been established for connections between hot-rolled steel sections in portal frames and in rectilinear frameworks used in multi-storey buildings. Details of the standard connections adopted in the UK are found in the publications by the SCI and the BCSA.[5,6] Their adoption makes fabrication more straightforward, promotes better communication, and reduces the chance of design and site errors.

Table 5.1 presents a summary of the preferred elements in standard connections. Fully threaded bolts have grown in popularity because they can be used with a wide range of thicknesses of the connected parts. However, they are not usually appropriate for exposed connections, unless the projecting length of the bolt is hidden.

Table 5.1 Preferred sizes of elements in standard connections

Element	Preferred option	Notes
Bolts	M20 Grade 8.8 bolts	Some heavily loaded connections may need larger diameter bolts Foundation bolts — M24 Grade 4.6 Fully threaded bolts may be used where the thickness of the connected parts is variable
Holes	22 mm diameter punched or drilled, or 22 mm x 26 mm slot punched	26 mm diameter for M24 bolts 6 mm oversize holes for foundation bolts
Welds	Fillet welds with E43 electrodes, 6 mm or 8 mm leg length	Larger welds may be needed for some column bases
Fittings	S275 steel Limited range of standard flats and angles	Refer to the SCI/BCSA publications[4,5] for further information

Some general detailing rules, as described in the SCI/BCSA publications,[5,6] are as follows. For nominally pin-jointed connections:

- the top of the top flange will generally be used as the setting out point
- the tops of all fittings (e.g. cleats) are placed 50 mm below the top of the beam
- the first bolt row is located at a constant 90 mm from the top of the beam, independent of the flange thickness
- the bolt rows are then set at 70 mm intervals below the row above. Therefore, the final row will be at a variable position above the bottom flange, depending on the section size.

For moment-resisting connections using end-plate type details:[6]

- 'flush' end-plates should be extended 15 mm above the top flange of the beam to allow for welding to the top flange of the beam
- 'extended' end-plates should be extended 90 mm above the top flange of the beam to accommodate an additional pair of bolts
- the first row of bolts below the flange is located at a constant 60 mm from the top of the beam, independent of the flange thickness
- the bolt rows are set at 90 mm intervals below the bolt row above, as for nominally pin-jointed connections
- for extended end-plates, the upper row of bolts should be located 40 mm above the top flange.

The following section deals with the typical range of connections for traditional applications.

5.4 Beam to column connections

The most common types of beam to columns used in buildings are described as follows:

- *Double angle web cleats*
  A double angle web cleat connection (Figure 5.1(a)) consists of a pair of angle cleats that are usually bolted to the beam web in the shop. The beam assembly is then bolted to the column on site. Single angles are much weaker as the bolts are loaded in single shear. The minimum size and thickness of angle section for a standard connection is 80 mm × 6 mm thick. The preferred size of angle is 90 mm × 10 mm thick for most applications.
- *Stool or seating cleats*
  A stool or seating cleat is sometimes placed under the end of the supporting beam, which provides a safe and positive landing position for the beam during erection. A web cleat is always used with a seating cleat in order to stabilise the member laterally (see Figure 5.1(b)).
- *Fin plates*
  Fin plates are most commonly associated with beam to beam connections, but may also be used for beam to column connections. Projecting plates may be welded to the column flange or web to which the incoming beam is bolted (see Figure 5.2).
- *Bracket connections*
  Other forms of bracket connections may be made to the side of columns. A good example is in the parallel beam approach, as illustrated in Figure 5.3 and as described in Section 4.2.5. Projecting channel sections are welded externally to the tips of the column flanges and extend outwards to connect to the beam web. The bolted connection through the end plate is made on site. This connection is not appropriate for small columns because of the difficulty of access for welding the bracket. The weld size is also determined by the torsional effects due to the continuous beam spanning over the main beams.

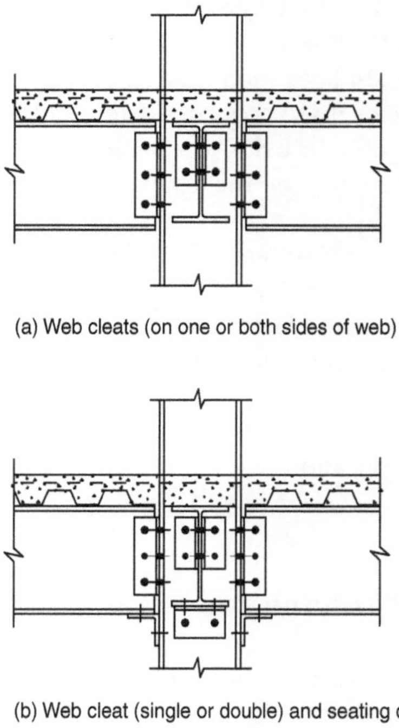

(a) Web cleats (on one or both sides of web)

(b) Web cleat (single or double) and seating cleat

5.1 Web cleated connections

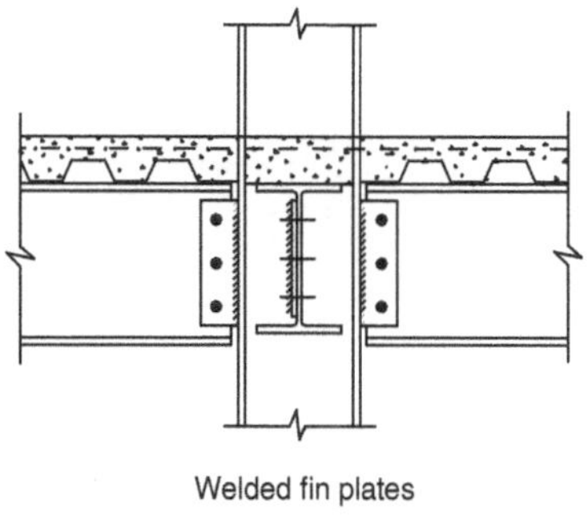

5.2 Fin plate connection (welded to column flange web)

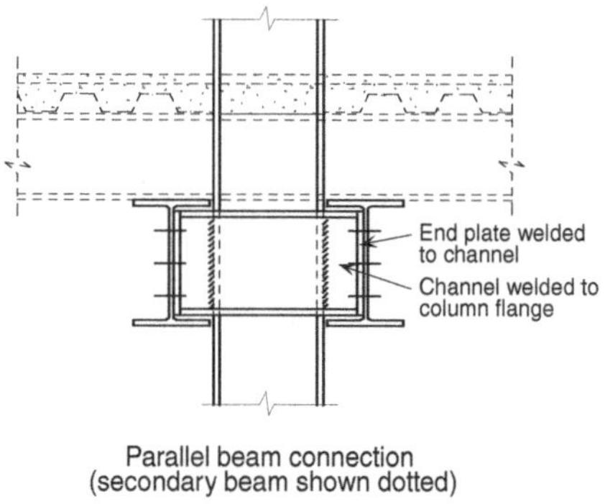

5.3 Welded bracket to connect pairs of beams

- *Flexible end-plates*
  'Flexible' end-plates consist of a thin plate welded to the beam in the workshop. The plate is typically 8 to 12 mm thick, depending on the size of bolts used. The beam is then bolted to the supporting member on site. End plates are probably the most popular of the beam-column connections used in the UK. They are versatile in that they can be used with skewed beams and can tolerate moderate offsets in beam to column alignments.
  They are termed 'flexible' because they are thin and are not necessarily welded to the beam flanges, and therefore do not transfer significant moments (see Figure 5.4(a)). The SCI/BCSA publication[5] gives guidance on typical details for these connections.
- *Welded shear blocks*
  Welded shear blocks, as shown in Figure 5.4(b), are widely used in continental Europe. The upper bolts are only used for location and to provide for tying forces. The welded shear block resists the vertical load transferred from the beam. However, the

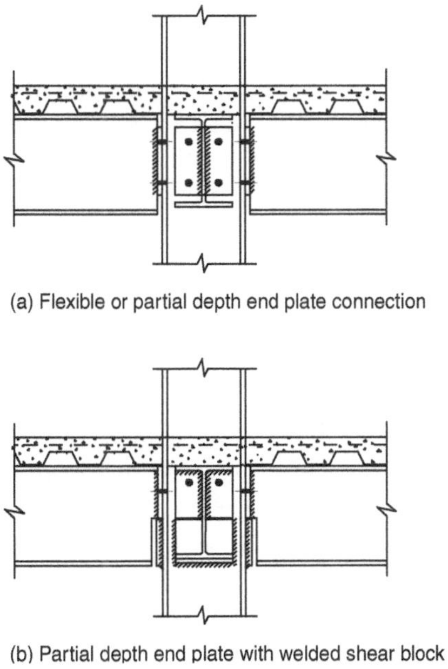

5.4 Flexible or partial depth end-plate connections

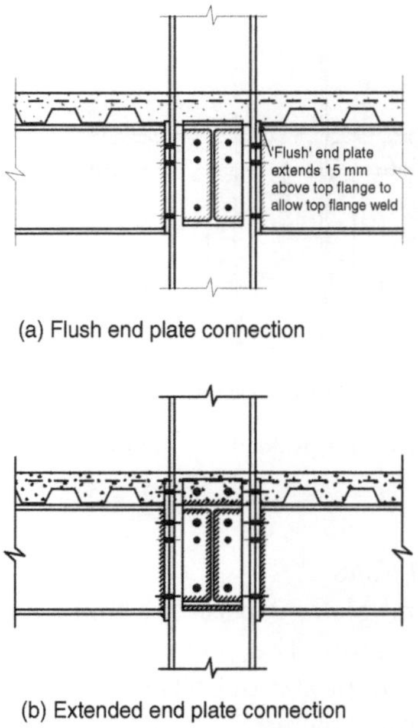

5.5 Moment resisting end-plate connections

block must be thick enough to allow for all tolerances in beam placement.

- *Thick end-plates*
  Moment-resisting connections between beam and columns can be fabricated by welding thicker end-plates to the beams. The end plate is typically 15 to 20 mm thick. Flush end-plates are welded to the flanges and web of the beam so that there is minimal (15 mm) projection of the end plate (see Figure 5.5(a)). Extended end-plates project above or below the beam depth (see Figure 5.5(b)) and achieve greater bending resistance by having the facility for bolts above and below the tension flange.

  The SCI/BCSA publication *Joints in Steel Construction — Moment Connections*[6] provides guidance on the practical application of these moment-resisting connections. Importantly, the number and size of the bolts is designed primarily to resist the tension forces caused by the applied moment. The lower bolts resist the applied shear-forces. Connections to the webs of columns require careful detailing, as the end plate to the beam must fit between the depth of the root radii of the column section. (The root radius can be up to 25 mm per flange.)
- *Haunched connection*
  A haunched connection, shown in Figure 5.6, is a further example of a moment connection that is typically used in single or multi-span portal frames. However, haunched composite beams have been used to create longer spans of minimum depth (see Section 4.2.3). The haunch is designed so that the connection is not the 'weak link' in the failure mechanism of the frame. It is usually created by cutting and welding a portion of the same beam-section in order to minimise wastage.
- *Welded connections*
  Fully welded connections are rarely used in building construction in the UK because of the potential difficulty in

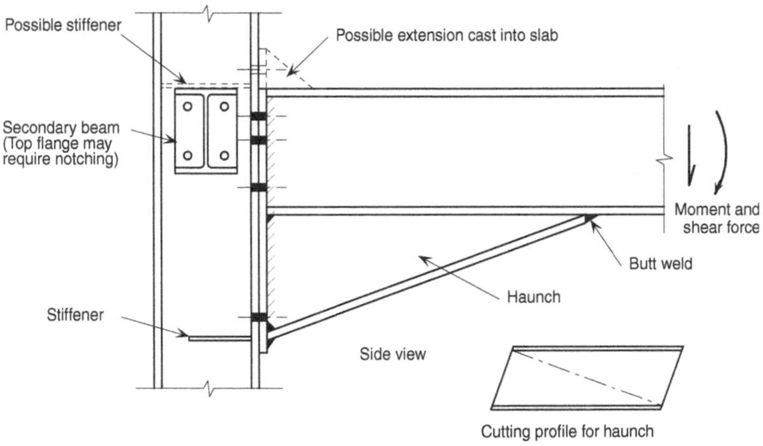

5.6 Haunched connection of a beam to a column

achieving good-quality welds on site. However, it may be possible to provide bolted splice connections elsewhere in the beam to facilitate transport and lifting, and to fully weld the main connections in the factory (as column 'trees').

- *Stiffened connections*
  Where the connection between the beam and column requires additional load-bearing capacity, or where the loading may be eccentric to the member axes, the connection may be stiffened in the form of welded plates of typically 6 to 12 mm thickness. Welding of stiffeners is relatively expensive and should be avoided in regular beam and column construction. It may be cost-effective to increase the column weight (size) to avoid the need for stiffeners.
- *Seated connections*
  In some low-rise buildings, seated connections may be used. Beams, or more usually, trusses are seated on end plates welded to the tops of the columns. Pairs of bolts provide for shear and uplift resistance. These connections are treated as pinned. Usually the supported members are restrained laterally by some other means, for example, a perimeter tie or eaves beam.

5.5 Beam to beam connections

The SCI/BCSA publication *Joints in Simple Construction*[5] provides guidance on the practical applications of beam to beam connections, which are generally treated as pinned. Practical conditions to be addressed are as follows:

- *Relative sizes of beams*
  A common feature of beam to beam connections is that the top flanges of the beams should be at the same level. Therefore, the ends of the secondary beams are often 'notched' so that they can be attached to the web of the primary beams (see below). Most of the previously described connection types may be used, and some are illustrated in Figure 5.7.

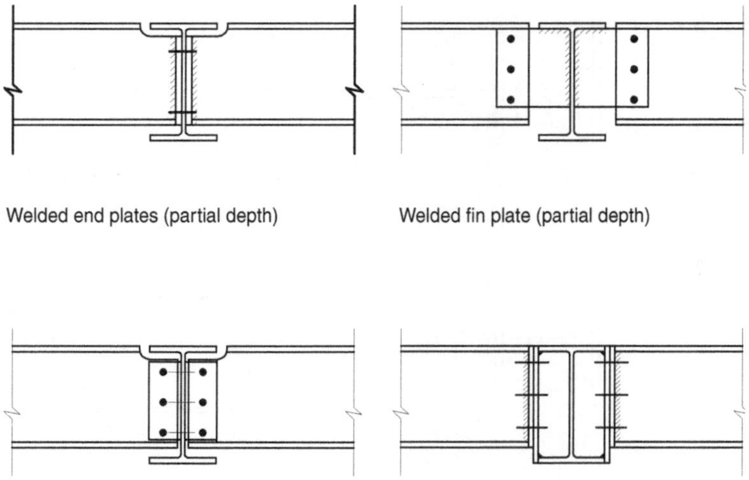

5.7 Typical beam to beam connections

Welded fin plates avoid the need for notching, but extend well outside the beam width and cause bending of the fin plate. Welded side-plates cause local bending of the flanges, and are not recommended for heavily loaded applications unless the web is stiffened to resist these local forces. A Tee section welded to the web provides this stiffening function.

- *Notching of beams*
  It is often necessary to notch or cut-back beams when connecting to other beams or columns. The detail in Figure 5.8 shows the amount of cutting back that is required in standard connections.
- *Splicing of beams*
  Spliced connections are rarely used in building construction except in very long-span beams where transportation or erection requirements necessitate the supply of shorter members. Spliced connections require web plates to transfer shear, and often top and bottom plates to transfer moment applied to the beam. Preferably, these splices are not made in the regions of high moment. Splice connections generally use high-strength friction grip bolts acting in shear to avoid the effects of bolt slip on deflections.
- *Connections of beams at different levels*
  In some building types, it is possible to align the beam at different levels, in which cases connections below the top flange may be made by end plates or web cleats, as in conventional connections. Beams suspended below beams may make use of special connectors, such as by Lindapter (see Figure 5.9). These connections may be of particular interest where drilling or welding is not permitted on site.

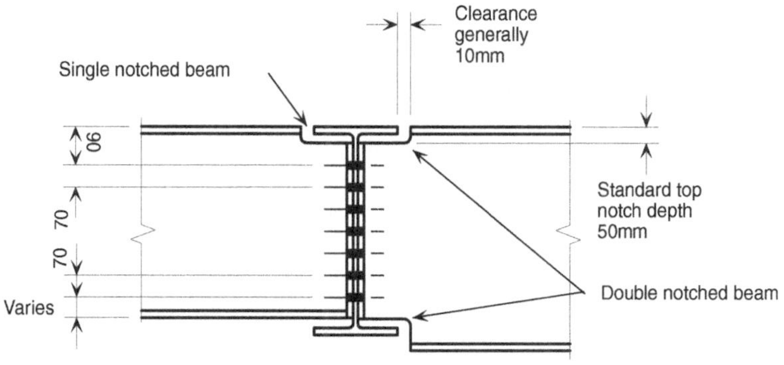

Standard notching of beam-beam connection

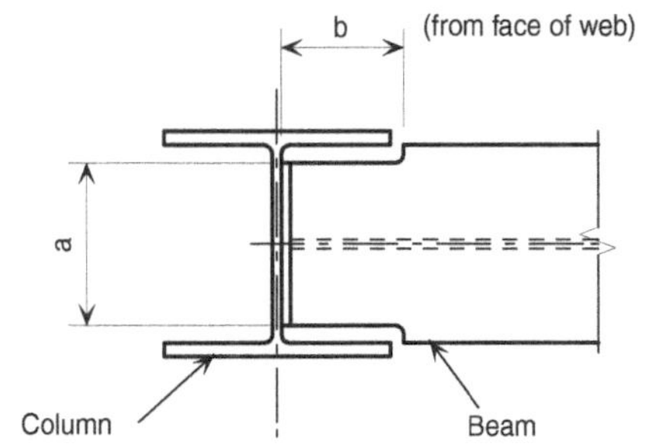

Plan view on beam-column connection

Column size	Flange trim (mm)	
	a	b
356 UC	240	190
305 UC	195	160
254 UC	150	135
203 UC	110	110
152 UC	70	85

Standard trimming for various column sizes

5.8 Typical detailing requirements for beam to beam connections

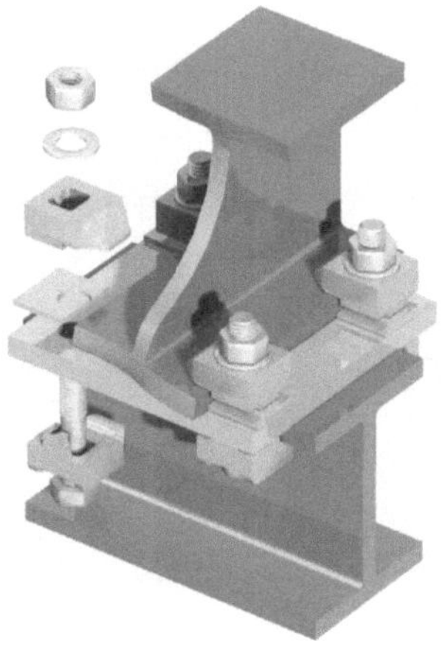

5.9 Suspended beam to beam connections by Lindapter

5.6 Column splices

Column splices in multi-storey construction are usually provided every two or three storeys, and are located about 500 mm above floor level. This results in convenient column lengths for fabrication, transport and erection. The splicing operation is safer and easier to perform if it is done at a reasonable working height. Section sizes for the upper levels can be reduced at splice positions, but the provision of splices at each floor level is seldom economic, since any saving in column weight is generally far outweighed by the additional costs of the fabrication and erection. Figure 5.10(a) illustrates typical column splices in columns of the same size, and Figure 5.10(b) illustrates splices at a change of column size.

There are two basic types of column splice: bearing and non-bearing. In the bearing type, the loads are transferred from the upper to lower columns directly, or through a division plate. To ensure efficient fit at the splices, the ends of the columns should be finished square. For lightly loaded columns, a sawn end is sufficiently accurate so that bearing surfaces do not have to be machined to achieve good contact. For larger, heavily loaded columns, the ends

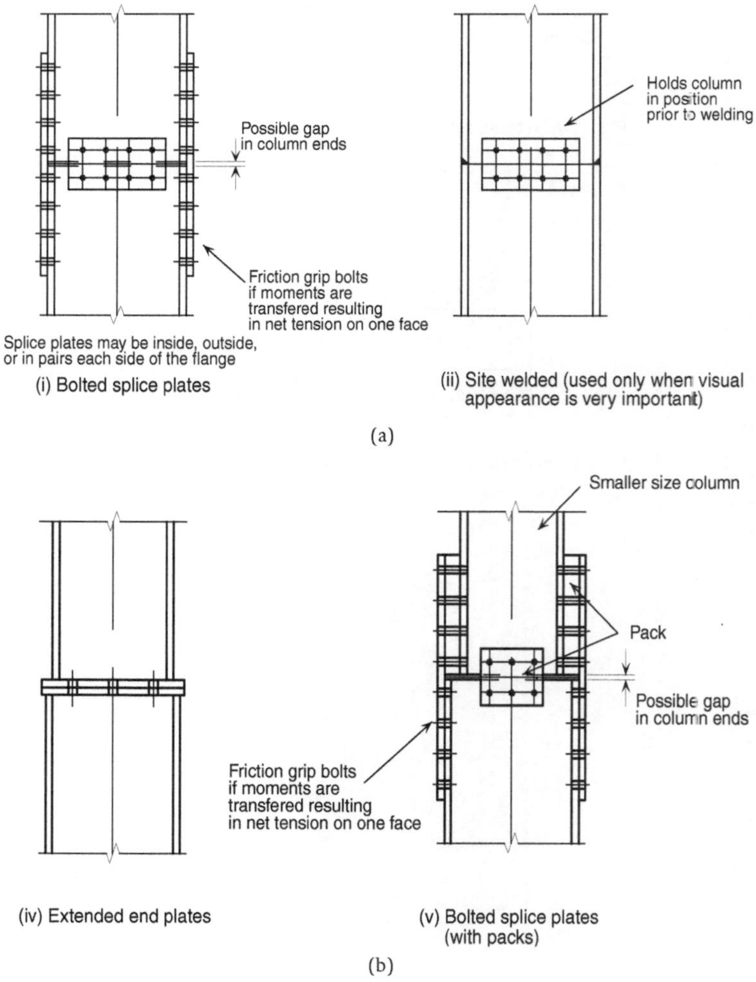

5.10 Various forms of column splices: (a) typical connections between column sections of the same size; and (b) typical connections between column sections of different sizes

should be machined in order to achieve good-bearing contact. In non-bearing splices, the loads are transferred by way of bolts and splice plates, and any bearing between the members is often ignored. Countersunk bolts may be used in the splice plates to avoid protruding bolt heads, which may otherwise interfere with finishes and fire protection, and may be less visually acceptable. However, this is a more expensive option than using conventional bolts.

5.7 Column bases

Column bases can be designed as nominally pinned (simple) or moment resisting (rigid). Nominally pinned bases are only required to transmit axial and shear forces into the foundation, and are provided in braced structures and in portal frames. They are generally preferred to moment-resisting base connections for reasons of cost and practicality. Uplift due to internal wind pressure and external wind suction may have to be considered in single-storey structures, which leads to a minimum size of foundation for a given building size.

Moment-resisting bases may be required in rigid-frame structures in order to reduce the effects of sway and deflections. These bases and their foundations are considerably larger than for nominally pinned column bases.

Holding down (HD) systems are designed to satisfy the following requirements:

- In service, they must transmit shear from the column to the foundation; if HD bolts are fitted using oversize holes in the base-plate, then shear must be resisted by other means.
- During erection, they must be capable of stabilising the column and other structural elements. Thus, four bolts are provided, even in a nominally pinned connection.
- They must resist uplift, depending on the design condition.

The base plate should be of sufficient size, stiffness and strength to transmit the compressive force and bending moment from the column to the foundation through the bedding material, without exceeding the local bearing capacity of the foundation.

Usually the force transfer from the column to the base plate is by direct bearing, and the welds between them are designed to resist shear only. Where required, the plate is designed for bending due to over-turning or uplift effects, which may cause tension in the HD bolts. Typical details for a column base using UC or SHS columns are shown in Figure 5.11.

Generally, the thickness of the base plate is chosen so that it does not require additional stiffening. However, there may be architectural merit in using shaped stiffeners in exposed applications.

To allow for tolerances in the concrete foundation, the top surface of the concrete is designed to be 30 to 50 mm below the bottom of the base plate. The column is temporarily supported on steel packs and wedges which permit vertical adjustment of the column. High-strength grout is then injected under the plate, and the wedges are

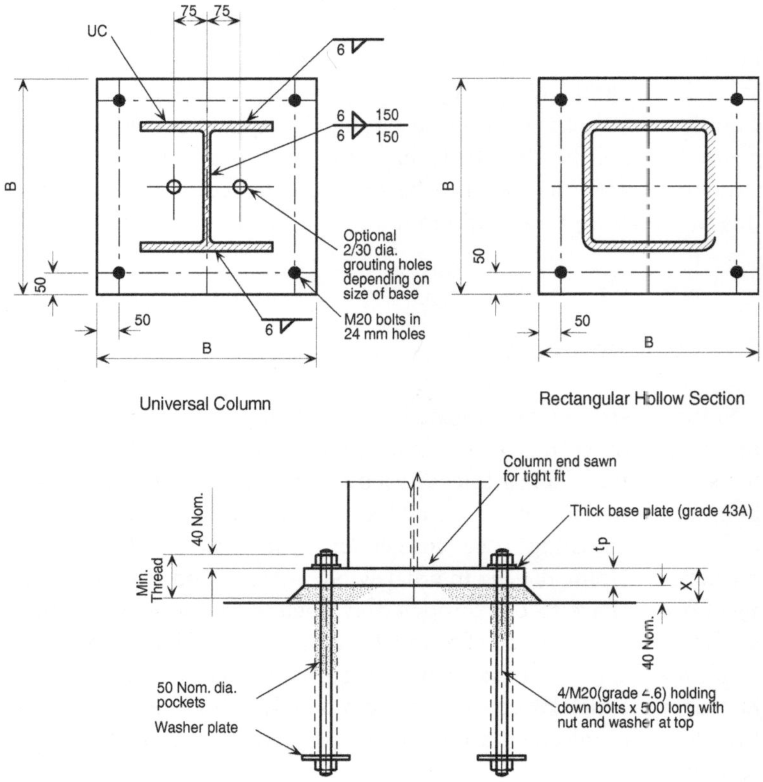

5.11 Typical simple column bases

removed when the grout has gained sufficient strength. Where column bases are required to be concealed, an allowance for this gap and for the end-plate and the projecting bolts must be made when determining the covering to this detail (typically 100 to 120 mm should be allowed). This may increase to 300 or 450 mm where rainwater downpipes are also located in the column zone.

5.8 Connections in trusses

Various forms of truss or lattice girder may be defined depending on the span and load configuration (see Section 4.5). Lattice girders have parallel top and bottom chords and are used as beams, whereas trusses may have inclined top chords for use in roofs. In both cases, the connections between the members may be bolted or welded. Welded connections are often preferred in tubular construction, or where the cumulative effect of bolt slip is critical to the design of the truss. Nevertheless, it may be necessary to introduce splices in the chord members if the trusses are too long for transportation. These splices should be located and detailed carefully if they are architecturally important.

5.8.1 Trusses comprising angle sections

Traditionally, roof trusses used angles, with bolted and gusseted connections (see Figure 5.12(a)). However, deeper T-sections for the

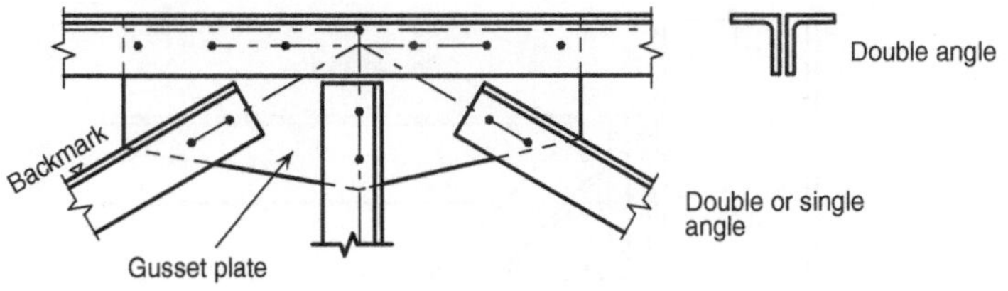

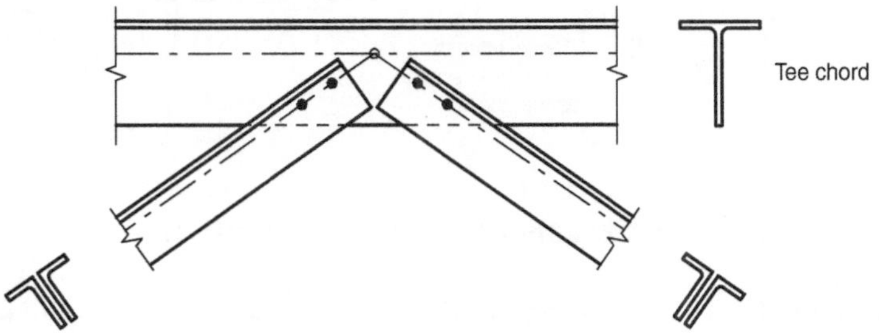

5.12 Traditional bolted connections in trusses

main chords avoid the use of gusset plates, provided the bolts can be accommodated (see Figure 5.12(b)). The projection lines of the bolt setting-out lines are detailed in such a way that eccentricities in the forces transmitted by the bolt groups are minimised.

In welded connections, the depth of the T-section is chosen so that the centroidal axis of the sections can be arranged to eliminate eccentricity (see Figure 5.13). Top and bottom chords are usually continuous, except at changes in direction or where splices are necessary for erection purposes. Pairs of angles, bolted or welded periodically along their length, are preferred, as they are much more resistant to buckling than single angles.

Lighter lattice girders used as secondary beams may be connected to continuous columns at their top chord only. This forms an effective 'pin' connection for design purposes (see Figure 5.14). However, heavier lattice girders supporting secondary beams should be connected to the columns at both their top and bottom chords.

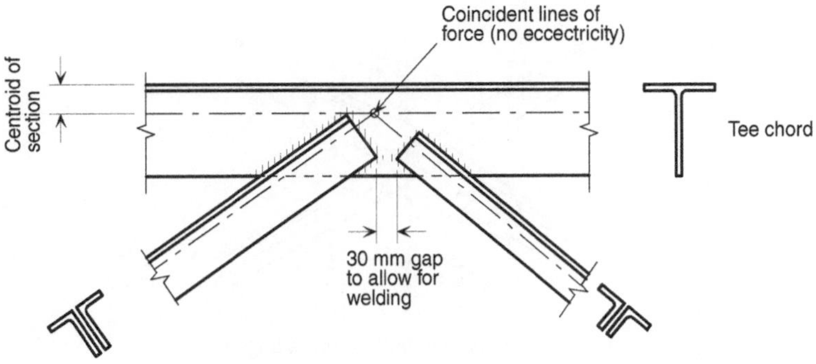

5.13 Typical bracing — chord welded connection

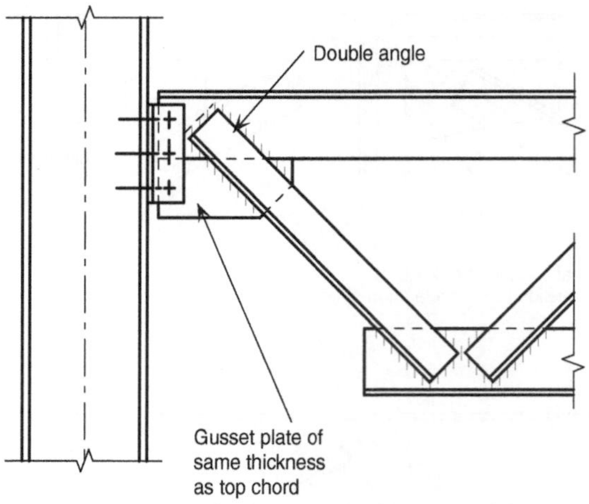

5.14 Typical truss-column flange connection

5.8.2 Lattice girders comprising heavier sections

Long-span lattice girders often comprise UC sections or tubular sections rather than angles in order to increase their compression resistance (see Figure 4.29). Heavy members may be required in special applications, such as transfer structures between floors which support point loads from columns above. In some cases, they are designed as storey-high assemblies. Deflection control is particularly important in long-span applications, and welded or friction grip bolted connections may be preferred to avoid the cumulative effects of bolt slip.

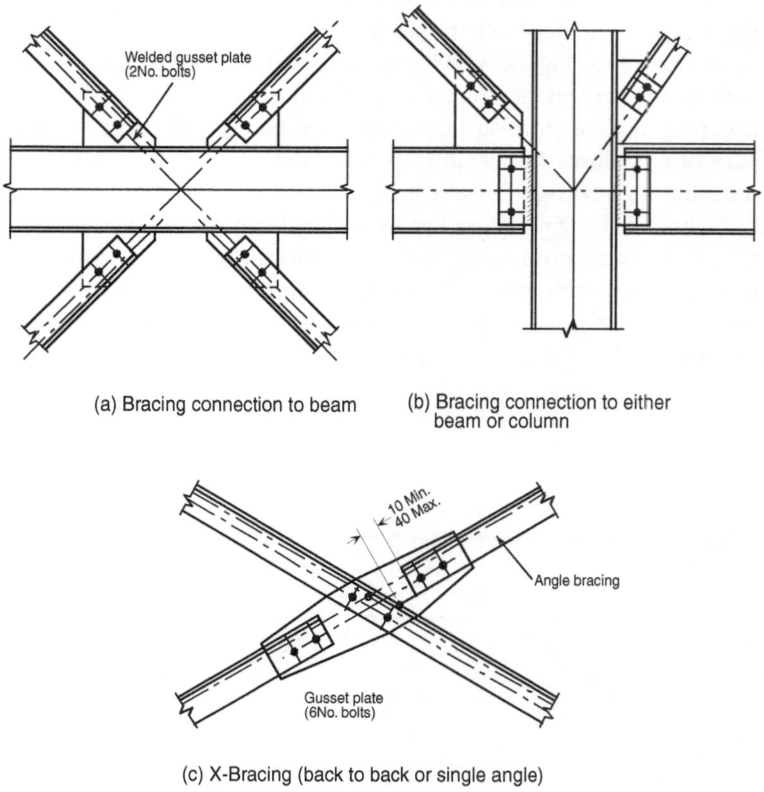

5.15 Examples of bracing connections in frames using angle sections

5.9 Bracing and tie-members

Vertical and horizontal bracing members resist wind and other horizontal loads applied to the building or structure, and transfer the loads to the foundations or other stabilising elements, e.g. concrete cores. In general, there are five forms of bracing and tie-members that may be considered: angles, flats, cables, rods and tubes. Some of them are only suitable for resisting tension, which dictates the form of construction in which they can be used.

The simplest form of bracing member is the steel angle, either placed singly or in pairs back to back. Single angles are less efficient in compression than double angles. Various forms of bracing assemblies may be used, such as X- and K-bracing, in which the members may be designed to resist tension or compression (see Section 3.6). Typical bracing connections using angle sections are shown in Figure 5.15. Angles designed for tension only will be more slender than those designed to resist both tension and compression.

Tubular connections are often preferred for bracing connections because of their good compression resistance (see Chapter 6).

Chapter 6

Connections between tubular sections

Connections between tubular sections are fundamentally different from those between open sections, such as I- and L-sections. There are two principal forms of connection: bolted and welded, but even bolted connections generally contain welded parts.

In designing tubular structures using circular (CHS), square (SHS) or rectangular hollow sections (RHS), it is important that the connection design is considered at the start of the design process, as this can have a strong influence on the aesthetics of the construction. This emphasises the need to consider the structural design and fabrication of the connections at the same stage as the selection of member sizes. Guidance on the use of tubes is given in various Corus publications (see Chapter 16).

6.1 Preparation of members

Tubular sections are connected by various techniques that require different forms of cutting and end preparation. The three techniques are:

- straight or inclined cutting
- end profiling
- end flattening.

6.1.1 Cutting

Straight or inclined cuts are made by conventional techniques and are used primarily for end-plate type connections. Shearing and punching of the sections is not recommended in heavy load-bearing applications. End plates are welded by fillet welds. Chamfering of the ends of the section is also possible in order to make a butt weld between tubes and to avoid a protruding end plate.

6.1.2 Profile shaping

'Profile shaping' is the process of shaping the ends of a tubular section to fit the contour of the curved surface of a tubular main member or chord to which it will be attached by welding at a suitable angle. Specialist fabricators can profile the ends of the section automatically by machine according to the required combination of member diameters and the angle between the centre-line of the members.

The profiled ends may also be chamfered at the same time to facilitate a butt-welded connection, which is neater externally. The connected members (which are generally inclined) may partly overlap (see later). This requires a different form of end profile on the overlapping member.

For small quantity production, hand flame cutting may be employed. Templates are used for marking off the section prior to cutting.

6.1.3 *Crimping or part-flattening*

CHS sections can also be flattened at their ends, if required. RHS sections cannot be flattened. End crimping of the connections in a space truss is illustrated in Figure 6.1. The taper of the flattened end should not exceed an angle of 1:4. These types of connection are normally used in small sections with two bolts placed through the crimped ends.

6.2 Bolted and pinned connections

Bolted connections in tubular construction are generally formed using one of the following techniques, as shown in Figure 6.2:

- welded end-plate with projecting end-plate or ring, so that bolts can be located externally to the hollow section

6.1 Use of an octagonal connector receiving flattened tubular struts, for a house at Almere, Holland (architect: Benthem and Crouwel)

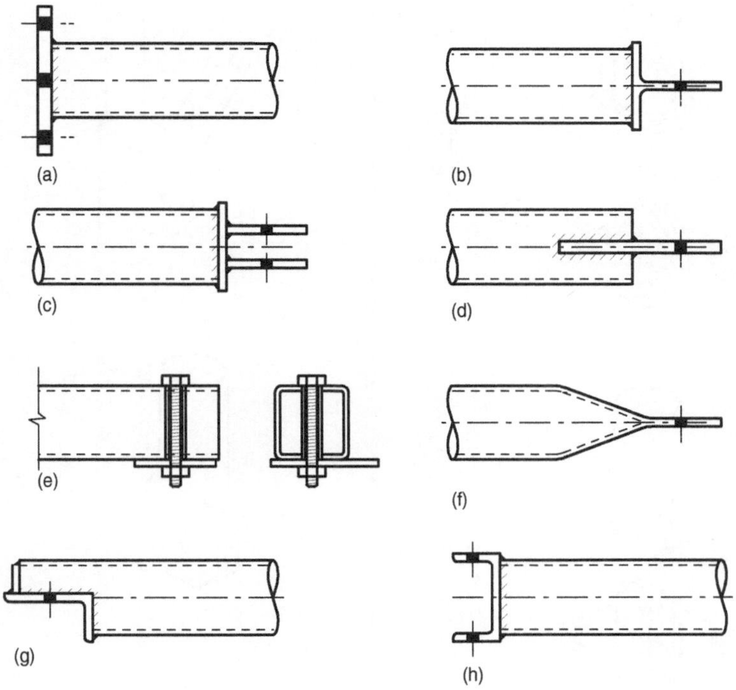

6.2 Some examples of bolted connections to tubular members

- welded plate with a projecting fin, often in the form of a T-section, which permits a conventional spliced bolted connection to be made
- welded fin that is welded to, or cut into the section, permitting a conventional spliced connection to be made
- through bolts or pins with internal ferrules to avoid local crushing of the walls of the hollow section
- flattened ends (of a CHS), permitting a spliced connection to be made
- welding to intermediary sections, such as angles or C-sections.

Bolted connections are desirable for site assembly, and large welded sub-assemblies that are prefabricated and bolted together on site at suitable locations. The practical aspects of installation should be considered in the design process. For example, Figure 6.3 shows

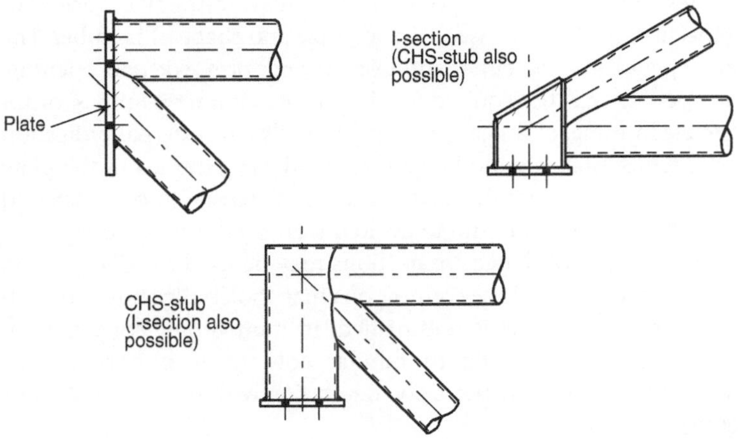

6.3 Bolted connections to the supports of tubular trusses

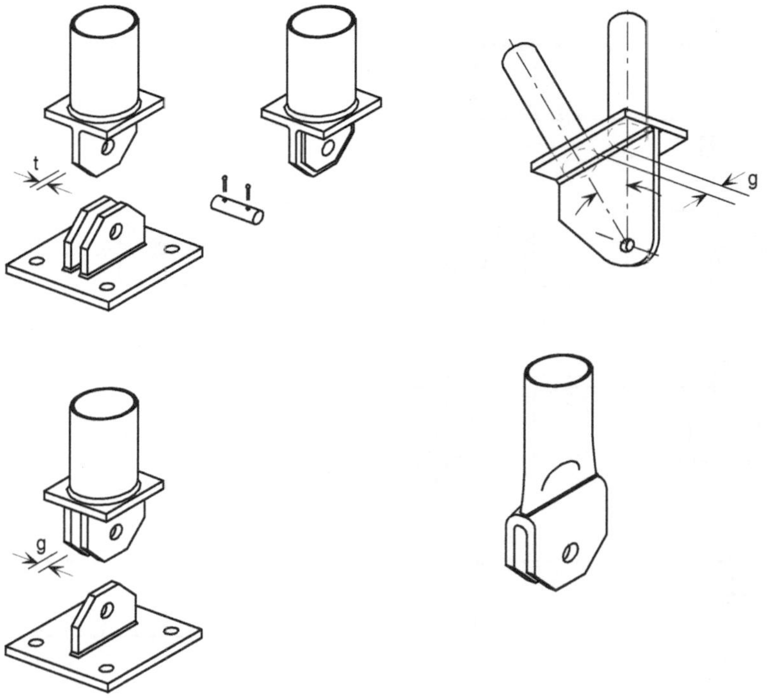

6.4 Some examples of tubular connections with pinned ends

possible end-connection details for long-span tubular trusses of various types.

Simple pinned connections can be made in a similar manner to bolted connections using welded end-plates and fin plates. Alternative pinned details for smaller tubular sections are shown in Figure 6.4.

6.3 Welded flange or end-plates and bolted connections

6.3.1 *Projecting flange-plates*

Welded flange-plates with projecting sides (see Figure 6.2 (a)) are the simplest but potentially one of the least attractive forms of connection, and can be used with any size and shape of member. The flange plates may be either solid or of 'ring-form' with an opening. The opening may be required for the passage of internal pipes, or for concrete infilling, or for galvanizing internally. The external projection of the flange plate should be kept as small as possible, but the plate should be of sufficient thickness to resist the tensile force transferred by the connecting bolts and to avoid distortion during welding.

Similar types of flange connections may be used for CHS or SHS sections. In close-up these connections may look bulky, but in overall perspective, their effect is diminished. In multi-storey construction, connections of this type to tubular columns can normally be accommodated within the floor depth (or within the raised floor depth).

6.3.2 *Welded plate with projecting fin plate*

This form of connection is an adaptation of the above type using a welded fin attached to the flange plate (see Figures 6.2(b) and 6.2(c)). The flange plate can be welded flush with the section by careful chamfering of the ends of the hollow section. The connecting bolts are then loaded principally in shear, as in a conventional splice connection.

6.5 Spliced connection between CHS and an I-beam

6.3.3 *Welded fin cut into the section*

A fin plate may be welded into a slot cut through or into the end of the section (see Figure 6.2(d)). In this application, the ends of the section may be sealed with a further semi-round plate or, in some cases, left partly exposed, where the risk of corrosion is small (i.e. in internal applications). The connecting bolts form part of a splice connection.

An interesting variant of this connection used to connect a CHS section to an I-beam is illustrated in Figure 6.5. Here the splice plate is curved at its end to enhance the visual effect. The four bolts transfer the required axial and shear forces.

6.3.4 *Through bolts with internal ferrules*

Bolts or solid pins may be passed through holes in the walls of the hollow section (see Figure 6.2(e)). Welded ferrules are located within the section to avoid local distortion on tightening of the bolts. These ferrules need only be tack-welded in place. The end of the section may be either capped or left exposed where the risk of corrosion is small.

6.3.5 *Sections with flattened ends*

Connections between sections with flattened ends are only appropriate in smaller CHS, such as in space trusses (see Figure 6.2(f)). Pairs of small diameter bolts are generally used in these splice connections which are often connected to prefabricated nodes or to similar sections with flattened ends.

6.3.6 *Welded fins*

Fins or brackets may be welded to the side of CHS or SHS/RHS sections to provide direct attachment of secondary members such as purlins (see Figure 6.6). Connections of this type require careful design because of the possible local distortion of the walls of larger hollow sections. Alternatively, welded threaded studs with extended washers may be used to attach the purlins to the section.

The attachment of tension-ties or rod-bracing members requires similar details. High local forces from ties may also be transferred by 'patch-type' connections, which may be profiled around the circular

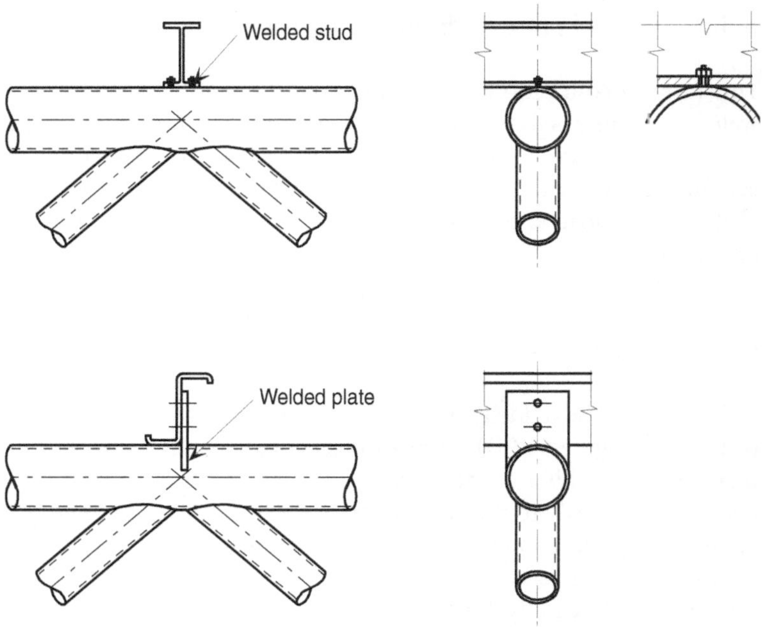

6.6 Bolted purlin connections to tubular trusses

6.7 Connection of ties to tube at Cologne Airport (architect: Murphy Jahn Architects)

section so that weld forces are transferred smoothly to the walls of the section. Multiple welded fin connections have been used successfully on a number of major projects, such as at the column bases at the Cologne Airport terminal, as shown in Figure 6.7.

6.4 In-line connections

Connections along the length of a member can be made by welding tubes end to end, or by a variety of bolted splices, as described below.

6.4.1 *Welded connections*

Welded in-line connections (see Figure 6.8(a)) are by far the neatest solution, particularly if the welds are ground back after fabrication. Welded connections can be designed to achieve the full strength of the tube, but they should normally only be made in the fabricator's shop in order to achieve correct alignment of the tubes. Changes in the thickness of the tubes can be accommodated at this point.

6.4.2 *Flange plates*

Flanged connections (see Figure 6.8(b)) are simple to make but are not aesthetically pleasing. They are suitable for compression but are less efficient for tension because of bending in the end plate, requiring thicker plates and more bolts. Fillet welding around the section could cause distortion of thin flange plates. Table 6.1 gives guidance on the flange-plate connections that achieve the full tensile-resistance of the given tube size. Fewer bolts or thinner plates may be used for lighter loadings and, in this case, the connection will be weaker than the tube.

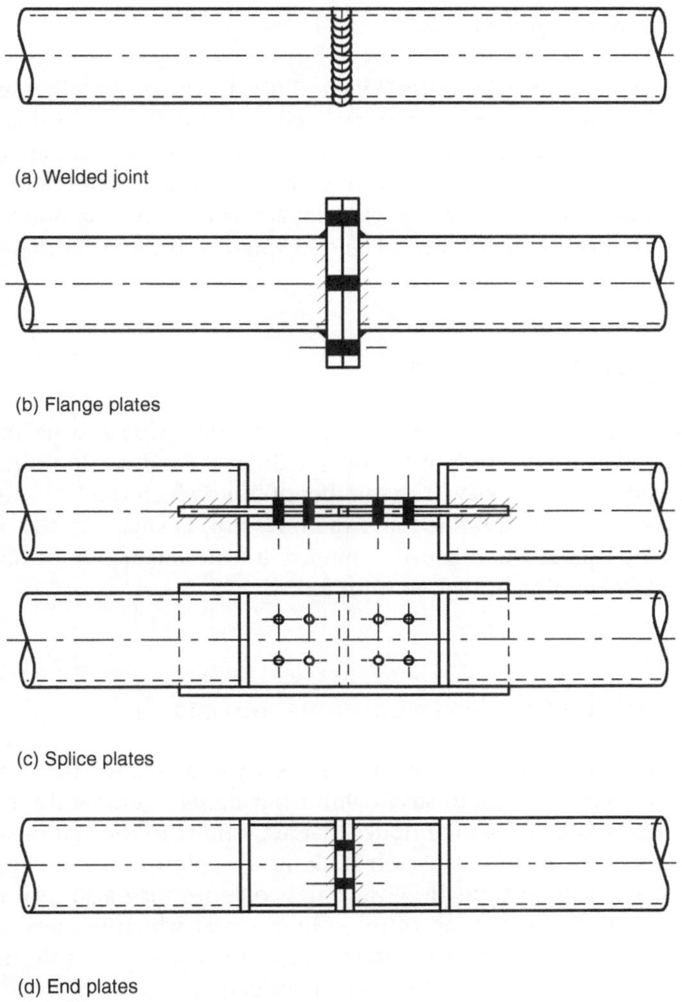

6.8 Examples of tube to tube splices

Table 6.1 Standard details for flanged connections (full-strength connections)

Max tube dimensions $d \times t$ (mm)	Thickness of flange plate t_f (mm)	Nominal diameter of bolt (mm)	Minimum no. of bolts	Edge distance (mm)
60.5 × 4.0 to 89.1 × 4.0	12	16	4	25
101.6 × 4.0 to 114.3 × 3.6	12	16	5	25
114.3 × 5.6 to 139.8 × 4.5	16	20	5	30
165.2 × 5.0	20	22	5	35
190.7 × 5.0	20	22	6	35
216.3 × 6.0	20	22	8	35
216.3 × 8.0	22	24	9	40
267.4 × 9.0	24	24	13	40
318.5 × 7.0	24	24	12	40
355.6 × 12.0	24	24	23	40
406.4 × 9.0	24	24	20	40

6.9 Splice connection

6.10 Connection with cover plate

6.4.3 *Splice plates*

Splice plate connections (see Figure 6.8(c)) can be made between tubes but the joint must be considered as pinned, making this an unsuitable connection for the middle of a member in bending or in compression. The splice plates can be either left exposed or used with a cover plate to give a smooth external appearance. An example of a splice connection is shown in Figure 6.9, and with its cover plate in Figure 6.10.

6.4.4 *Partial end-plates*

Figure 6.8(d) shows the detail of a very neat joint using a partial end- or side-plate, particularly if the open side of the connection can be arranged away from view. The number of bolts which can be located inside the section is limited and the lever arm is small, so that the connection should be regarded as pinned. It is unsuitable for members subject to bending or high-tension forces.

6.5 Welded nodes to columns and masts

In tension-tie structures, it is often necessary to attach the ties in the form of rods or cables to steel columns or masts. Multiple ties may be connected by fabricated nodes that are welded to the columns at the top, or at intermediate points along the columns, as shown in Figure 1.2. Connections in tension-tie structures may also take the form of saddles at the top of the columns over which the ties run. These saddles are fabricated and welded to the ends of the columns.

A striking example of the use of welded nodes is the tension structure at Darling Harbour, Sydney (see Figure 6.11). Four columns are placed together and the nodes are grouped in order to connect to the ties which support the long-span trusses. Another example of the innovative use of column clusters is shown in Colour Plate 21.

The 30 St Mary Axe building is constructed using a diagrid of intersecting tubular members with welded steel nodes acting largely in compression (Colour Plate 1). The nodes also support the perimeter ties and the internal beams, and an example is shown in Figure 6.12.

6.6 Pinned connections to tubular sections

Tubular connections provide the opportunity for true 'expressed' pins, as follows.

6.6.1 *Column bases*

Bases to tubular columns take two basic forms: pinned and rigid (or moment-resisting). The details employed reflect the transfer of forces and moments. A genuine pinned connection can be achieved by a single pin from a projecting plate, as shown in Figure 6.13. A

6.11 Column cluster arrangement at Darling Harbour, Sydney (architect: Philip Cox & Partners)

6.12 Welded node of the 30 St Mary Axe building — see also Colour Plate 1 (architect: Foster and Partners)

moment-resisting connection is achieved by a welded end-plate with four or more bolts. The thickness of the end plate depends on the moment to be transferred (see Section 5.7).

6.6.2 *Expressed pinned connections*

6.13 Typical pinned connection to a foundation

Connections using true 'pins' provide much scope for the literal interpretation of a rotationally flexible connection between members in a 'pin-jointed' assembly. Pins are usually made from two or three components. A central pin connects two ends or heads by passing through a hole in the connecting plates. The pin can be made from mild or stainless steel, and is generally smooth internally and threaded at its ends. If it is made from stainless steel, neoprene washers must be inserted to prevent bimetallic corrosion taking place with any attached mild steel elements. True pinned connections are shown in Figure 6.13 and Figure 6.14. Interesting details can be created using cast iron or cast steel nodes in a pinned connection.

6.14 Pinned connections at Ponds Forge Swimming Pool, Sheffield — see also Colour Plate 20 (architect: FaulknerBrowns)

6.7 Welded tube to tube connections

The form of welded connections between tubular members depends on the:

- shape and relative size of the members to the connections
- angle of intersection of the members
- number of members to be connected at one location.

Some fabricators are specialists in tubular construction and can advise on costs and details at the planning stage. Additional aspects, such as the need for the grinding of welds and any special connection details should be identified at this stage.

In terms of fabrication cost, a lattice girder using CHSs would require about 30 to 45 hours' work per tonne, and a similar lattice girder of triangular cross-section would require about 70 hours' work per tonne. When using larger CHS, fabricators with specialist profiling equipment can make the connections between the chords and web members efficiently. The alternative may be to use SHS sections, which only require cutting the ends of the chord members at the correct angle rather than profiling the cut ends.

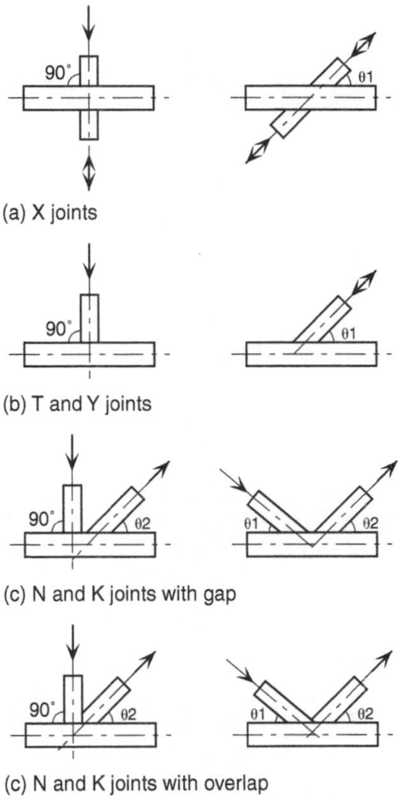

6.15 Connection designations in welded tubular construction

6.7.1 Typical welded connection configurations

Welded connections which are standard throughout the industry are known as X joints, T and/or Y joints, N and/or K joints, with or without overlaps, as illustrated in Figure 6.15. The precise form of these connections depends on the size and shape of the members. Gaps or overlaps between the bracing or incoming members can be detailed, and influence the load capacity of the connection (see Section 6.8).

6.7.2 Connections between square or rectangular sections

Welded connections may be formed relatively easily between the ends of one member and the flat wall of a larger SHS or RHS section. The structural engineer should check the local capacity due to distortion of the wall of the main member or chord when section sizes differ considerably and high forces are to be transferred. It should be noted that the resistance of the connection will be dependent on the size of the members rather than the strength of the weld.

The connection design should therefore be carried out at an early stage to avoid costly and potentially unsightly changes at a later stage in the design process. Welds may be formed by fillet welds externally, or by partial penetration welds to the prepared ends of the incoming section. The second detail is more attractive visually. Welds may be ground down where visually important.

Bracing members are generally aligned so that the centre-lines of the bracing members meet at the centre-line of the main chord in order to minimise secondary bending effects in these members.

The minimum angle of intersection of SHS or RHS members for welding is 30° to the axis of the main member, although, in practice, these connections should be made at an angle close to 45°, so that access for welding is less difficult.

6.7.3 Connections between circular sections

Welded connections between CHS require careful cutting and preparation to form the correct profile at the end of the incoming member. Profiling should also take account of the location and size of other intersection members. Severely overlapping member-connections increases the difficulty of profiling and welding. The minimum angle of intersection of CHS members for welding is 20° to the axis of the main member. Advice should be sought regarding welding of different sizes of members at shallow angles.

Three-dimensional welded nodes can be extremely complex, as seen in offshore construction. These nodes may be prefabricated, and the chord and bracing members are welded to the prefabricated nodes. It may be economic to consider the use of prefabricated cast steel nodes where the repetition of details can be achieved.

6.8 Connections in trusses and lattice construction

6.8.1 Two-dimensional trusses

Tubular sections are commonly used in long-span trusses for reasons of aesthetics and structural efficiency. Generally, CHS members are used for both the chords and bracing members, and a typical welded connection is illustrated in Figure 6.16. However, the top and bottom chords may use RHS rather than CHS members in order to facilitate

6.16 Welded connection of CHS bracing members to CHS chord

the connection with the roof or floor slab and other cross-members (an example of this type of detail is shown in Figure 6.16).

Lattice trusses have traditionally been designed as pin-jointed assemblies in which the members are in tension or compression and the forces between them are transferred at the connections. It is usual practice to arrange the connections so that the centre-lines of the bracing members (branches) intersect on the centre-line of the main member (chords), as shown in Figure 6.17. This is known as 'noding'.

Whilst 'noding' is common practice, for ease of fabrication it is sometimes required to provide a small degree of eccentricity of the nodes (as illustrated in Figure 6.18). A node with negative eccentricity may be architecturally more interesting, although a node with a total overlap is less so. The structural engineer or steel fabricator will advise on specific details.

Other connections between the elements of a truss can be made in various ways. Figure 6.19 shows various forms of right-angle connection at the end of the truss. Figure 6.20 shows connections of the inclined bracing members to the bottom chord). An example of the above detail is shown in Figure 6.21

Simpler connections in shorter span trusses can be made by bolted connections using gusset plates welded to the main member or chord. A simple detail in which the CHS bracing members have flattened ends is shown in Figure 6.22. A more architectural example of a bolted splice connection with a curved gusset-plate is illustrated in Figure 6.23.

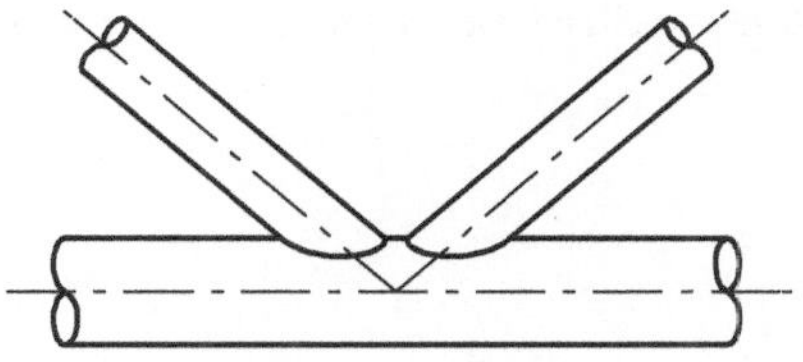

Gap joint noding

6.17 Illustration of the alignment of centre-lines of tubular members in a welded connection

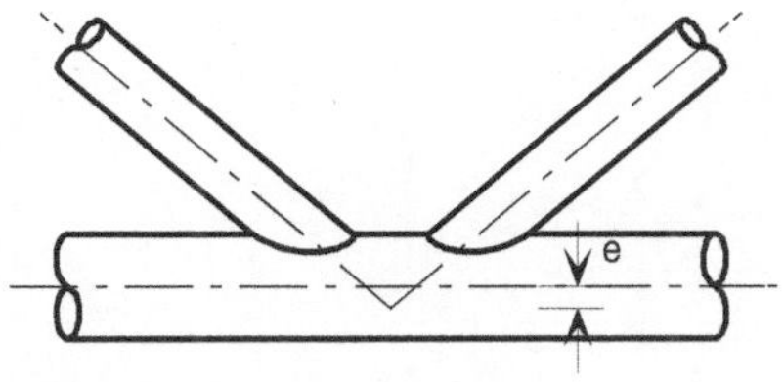

(a) Gap joint with positive eccentricity

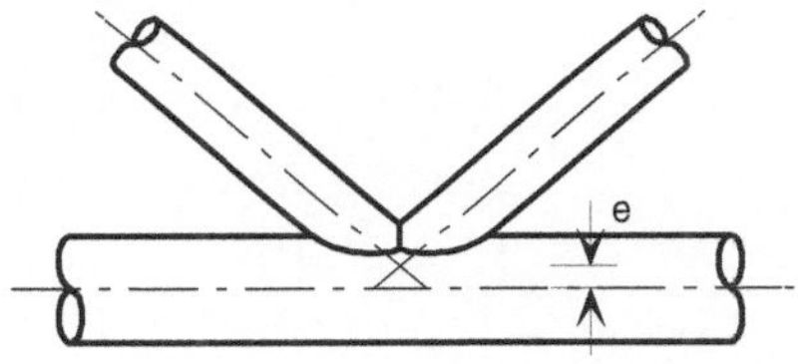

(b) Partial overlap with negative eccentricity

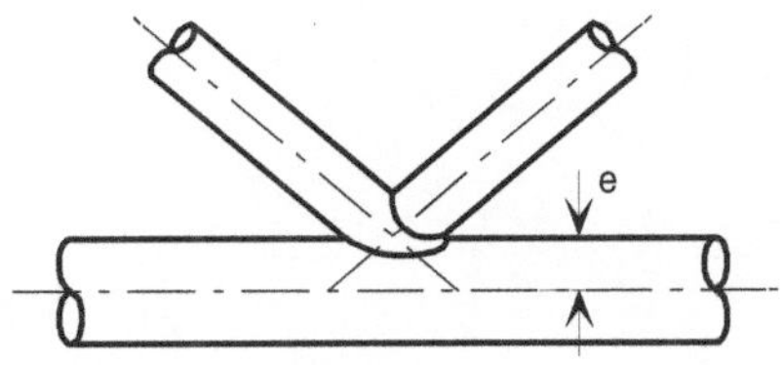

(c) Total overlap joint with negative eccentricity

6.18 Examples of noding with modest eccentricity

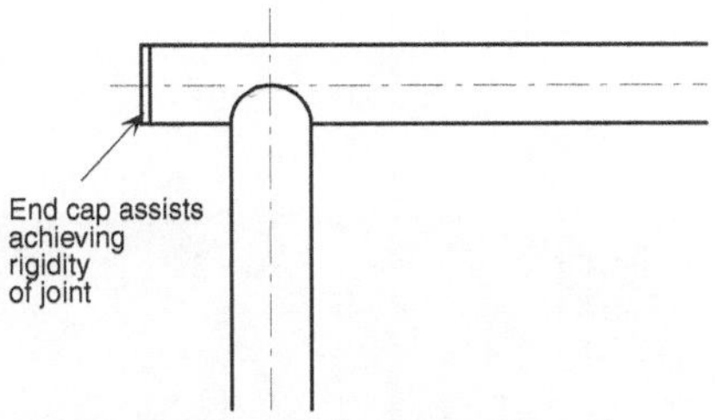

(a) Overshooting right-angle connection

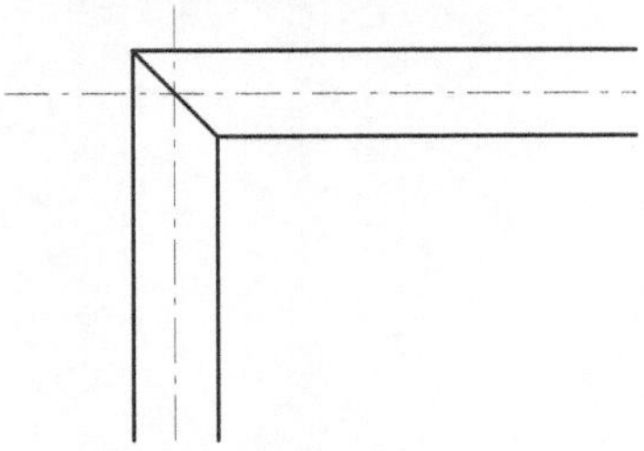

(b) Flush right-angle connection

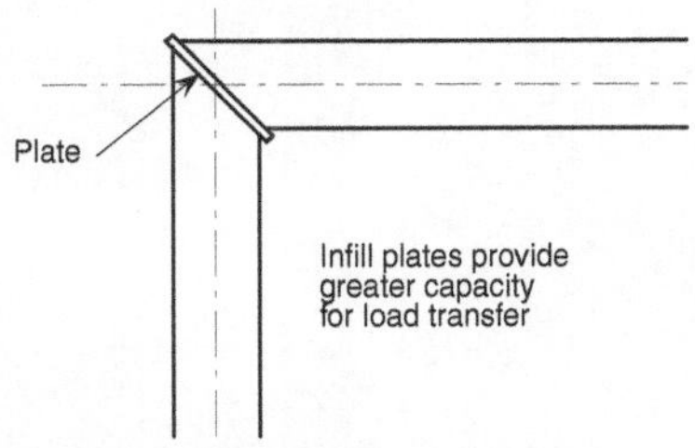

(c) Flush right-angle connection with infill plate

6.19 Right-angle connections between tubular members

6.20 Inclined connections in a lattice truss

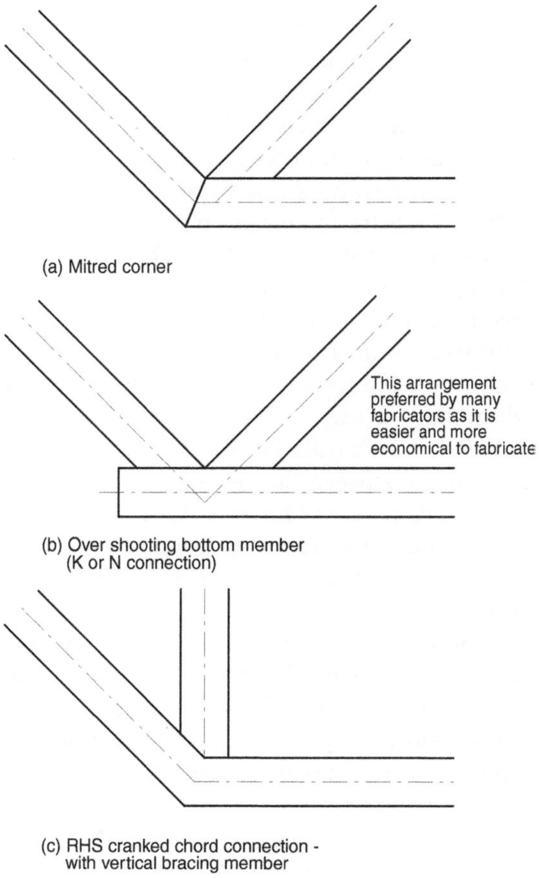

6.21 Example of welded CHS connection in a truss

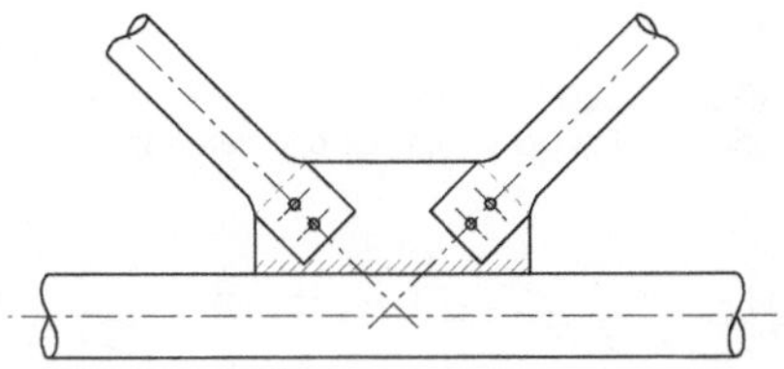

6.22 Gusset-plate connection

6.23 Architectural use of a bolted gusset-plate connection in a lattice truss for a railway bridge

6.8.2 Connections in multi-planar trusses

Trusses can also be designed in triangular cross-section along their length, as shown in Figures 6.24 and 4.42. These triangular section trusses have several advantages over plane trusses, because of:

- the increased stability offered by the twin compression chords — they are frequently used as exposed structures with long spans
- the simplification of bracing requirements in roof structures, in which in-plane forces have to be transferred along the roof
- their ability to resist torsional effects from incoming beams or trusses.

In this form of construction, there are various possibilities for the alignment of the chord and bracing members, as shown in Figure 6.25. Overlaps of the intersecting bracings from both planes may occur where the chords are smaller in diameter than 1.4 × bracing member diameter. This may occur in an offset connection as shown in Figure 6.25(c). Where many members come together at one node, this is known as a 'multi-planar' connection.

Two alternative bracing arrangements in triangular lattice trusses are illustrated in Figure 6.25. The configuration in Figure 6.26(a) requires more complex welding of the nodes at the bottom chord. The simpler connection detail in Figure 6.26(b) facilitates welding by arranging for a greater symmetry in the bracing arrangement. For

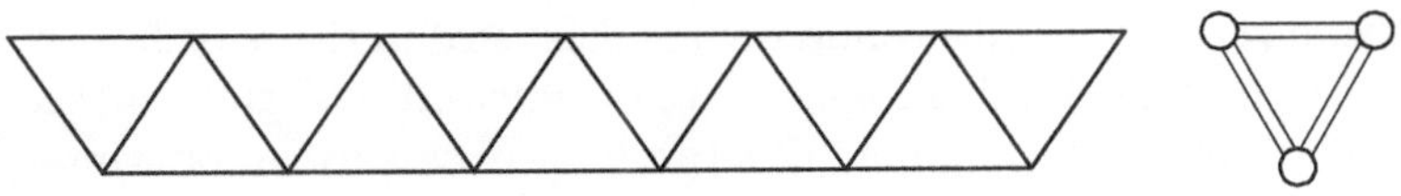

6.24 Triangular truss in cross-section

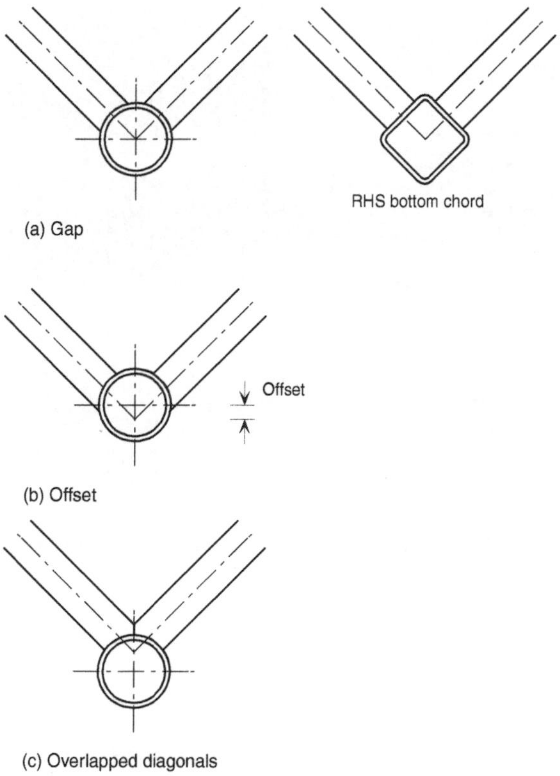

6.25 Connection types used in triangular section trusses

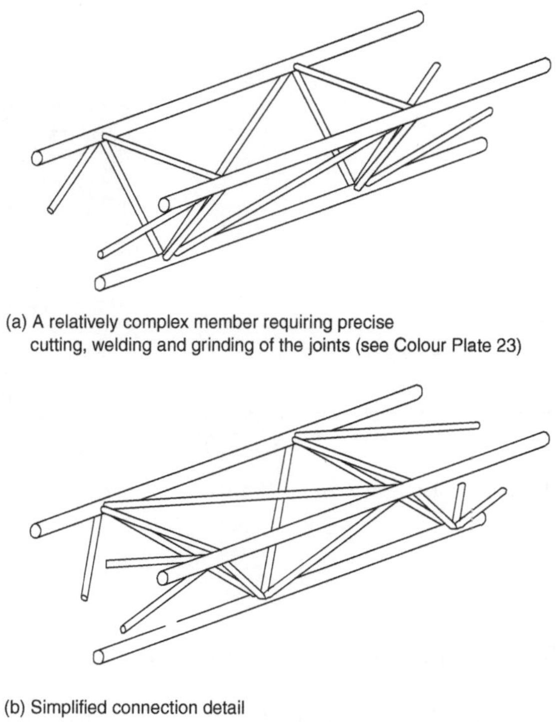

6.26 Alternative bracing patterns for triangular lattice girders

RHS chords, it may be necessary to increase the wall thickness to provide more resistance to forces transferred from the bracing members.

6.8.3 Reinforcement of connections

For maximum resistance of the members, it is usually more efficient to select larger tubular sections with thin walls. However, when designing the connections, it is more advantageous to use chord members that are thicker and smaller in section (provided that they are not smaller than the bracing members). Therefore, a compromise is necessary for overall design and fabrication efficiency.

In some cases, connections may have to be strengthened locally to resist the applied forces, if it is not possible to increase the member size or thickness. This can be achieved by welding plates to the chord face (see Figure 6.27(a)). It should also be noted that overlaps will also increase the connection resistance, especially for RHS members. When a third member is required at the intersection, a 'T' piece can also be used (see Figure 6.27(b)).

Other non-standard stiffened K connections can be used to increase the load capacity of the connection, as illustrated in Figure 6.28.

For multiple-bracing connections, the intersections can be moved back from the node point. This can be achieved by introducing a

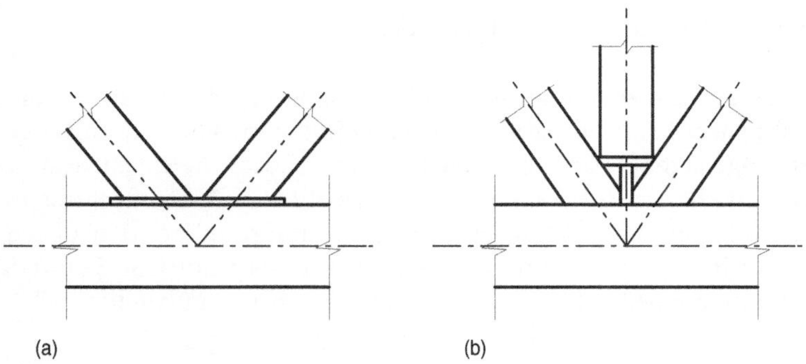

6.27 (a) Adding plate to chord section; and (b) adding a T-plate to facilitate connection

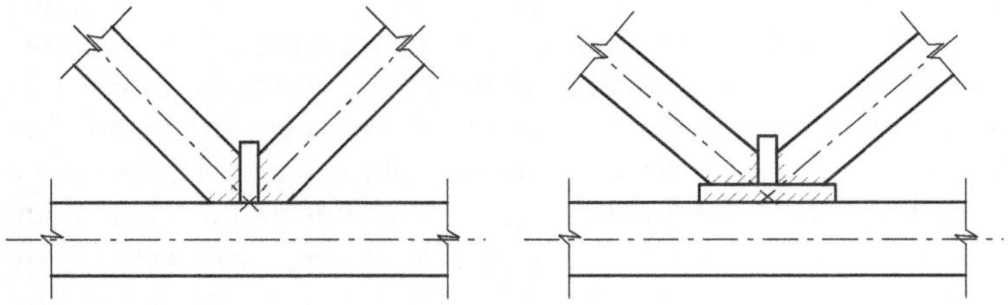

6.28 Additional stiffening plates to create non-standard K connections

short length of CHS, or by employing hollow spheres. Spheres have the advantage of the same cut at the end of the member, connecting the member's end regardless of the intersection angle. However, the source of this type of node is limited.

Saddle reinforcement can be used to locally increase the chord thickness and local compression resistance, as illustrated in Figure 6.29.

At the headquarters of Royal Life in Peterborough, one of the features is a glazed-screen elevation sweeping in a curve from one block to another. The façade was located 700 mm away from the primary structure, and it required its own support structure. The designers increased the stiffness of the section by welding four pairs of longitudinal steel fins, which, in turn, match the metal fins of the cladding (see Figure 6.30).

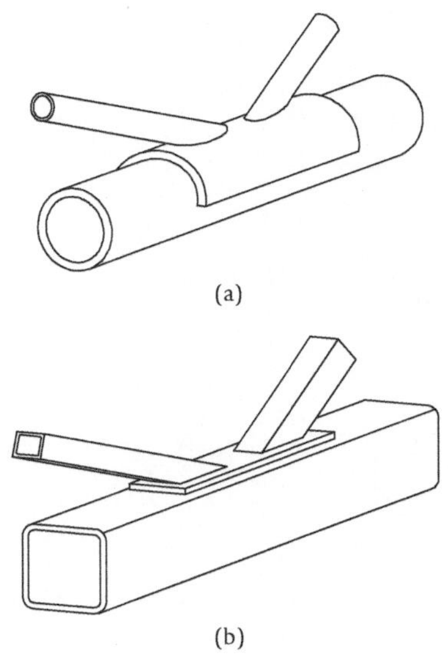

6.29 Reinforcement to tubular sections to increase their local resistance to forces from the bracing members: (a) saddle reinforcement; and (b) flange plate reinforcement

6.30 Royal Life UK headquarters — steel tubular column with four pairs of fins (architect: Arup Associates)

6.8.4 *Connections in Vierendeel trusses*

Vierendeel trusses comprise members connected at right angles and resist shear loads primarily by bending in the members. In this way, bracing members are eliminated but the chords are much heavier because they resist bending as well as axial forces. Vierendeel trusses employ only rigid or full-moment connections, unlike triangulated trusses in which the connections are designed as pinned. SHS or RHS sections are generally used in Vierendeel trusses, rather than CHS sections, because of their better bending resistance.

Vierendeel trusses are relatively inefficient at resisting high shear-forces because of the lack of diagonal bracing and, therefore, it is necessary to use thicker or larger chord members than in triangulated trusses. Ideally, the chord and vertical members should be the same external size. If not, stiffening elements are generally inserted to increase the local bending resistance of the connections. Figure 6.31 shows various ways in which nominally pinned connections can be strengthened in Vierendeel trusses. Visually, some of these details are not preferred, except when the trusses or these connections are hidden. An example of a 'hybrid' welded and bolted connection between RHS sections is shown in Figure 6.32.

6.9 Beam to column connections in tubular construction

The configuration of beam to column connections depends on the type and size of members to be joined. Three generic types of connection exist:

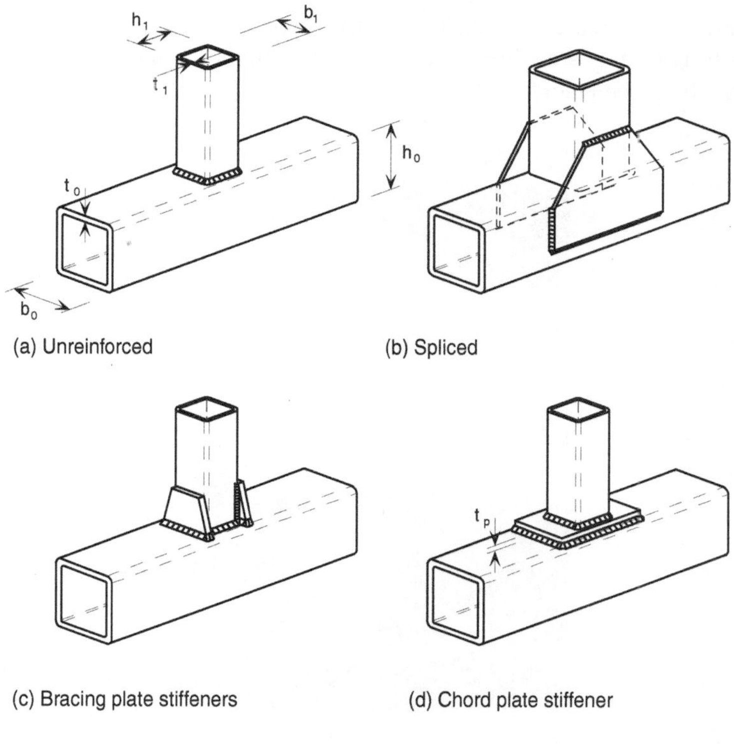

6.31 Types of Vierendeel connections between SHS/RHS members

6.32 'Hybrid' welded and bolted connection

- RHS beams to I-section columns
- I-section beams to CHS or SHS columns
- RHS beams to CHS or SHS columns.

Beams and columns are usually connected on site by bolting. In the case of an RHS beam connection to an I-section column, a welded extended end-plate to the RHS beam permits the use of a conventional bolted connection to the column flange or web (see Figure 6.33). The bolts may be countersunk into the thick end-plate if the connection is important visually.

In the case of bolted connections to SHS or RHS columns, special forms of bolts are required, which can be located from one side. *'Flowdrill'* and *'Hollo-Bolt'* are two particular forms of bolt suitable for use with SHS or RHS sections (see Section 6.10). Alternatively, brackets or fins can be welded to the RHS column to provide direct bolted connections.

A number of typical simple connections using cleats welded to an RHS column are shown in Figures 6.34 to 6.36. Figure 6.34(a) shows a fin plate welded to the face of the column. The supporting bracket in Figure 6.34(b) can be detailed to be visually interesting. Figure 6.35 shows the use of channels welded at the tips of their flanges to

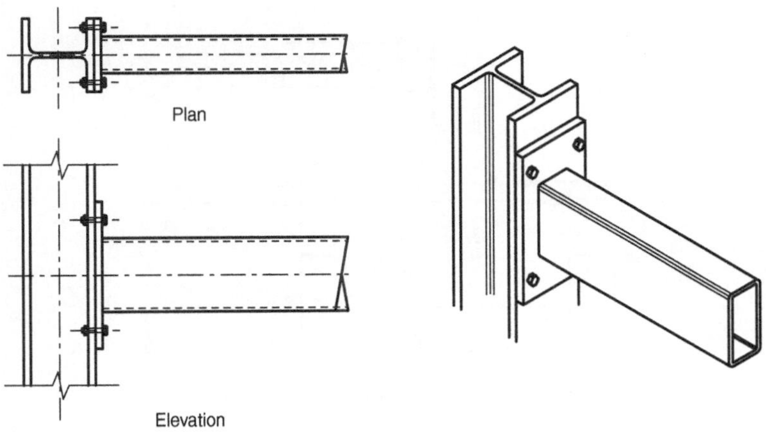

6.33 Connection of RHS beam to I-section column

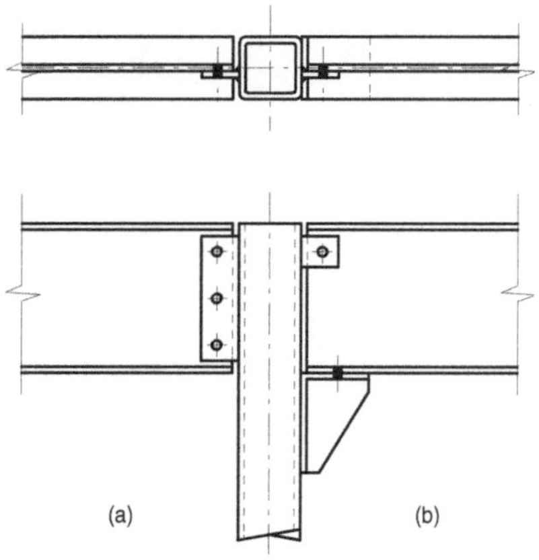

6.34 Conventional cleats welded to an RHS column

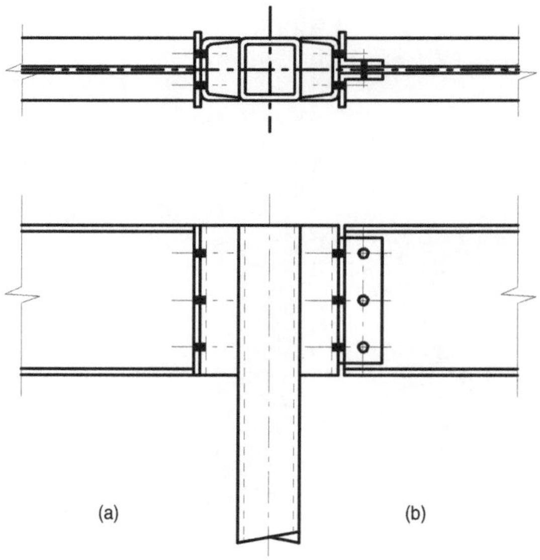

6.35 C-sections welded to an RHS column to facilitate the use of a bolted connection

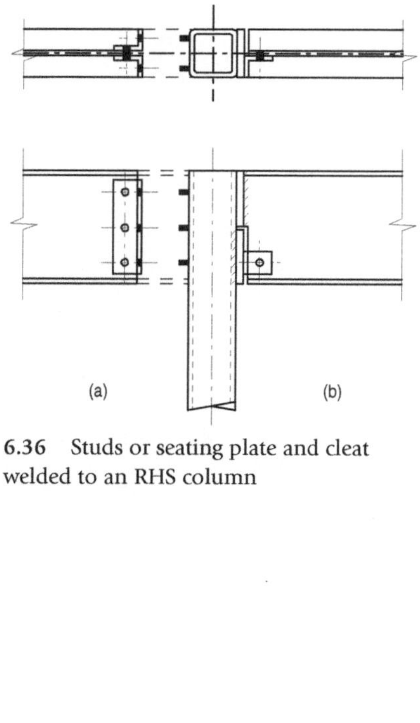

6.36 Studs or seating plate and cleat welded to an RHS column

permit access for bolting on site. Figure 6.36(a) shows the use of welded threaded studs, but these must be protected during transit to prevent damage. Figure 6.36(b) shows an I-section with a partial depth end-plate connection supported by a shear block welded to the column. The welded block must be sufficiently thick to allow for all site tolerances. Also, the single bolted connection may not be acceptable for 'robustness' requirements in multi-storey buildings.

Figure 6.37 shows other typical connections of an RHS beam to an RHS column. For lightly loaded connections, the T-section shown in Figure 6.37(a) may be replaced with a fin plate. Where through bolting is used (as in Figure 6.37(b) and 6.37(c)), spacer tubes

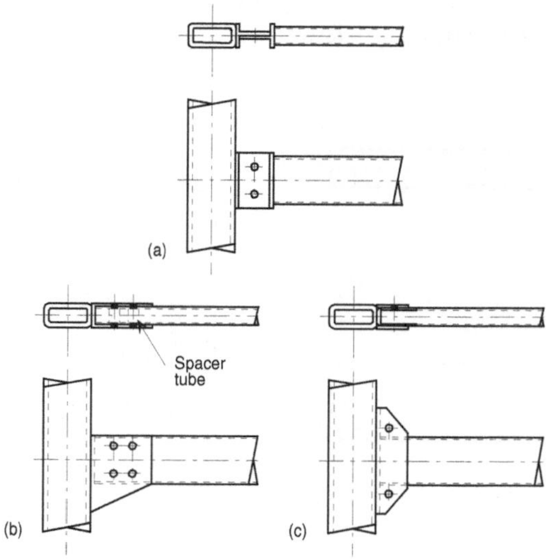

6.37 Details of RHS beams connected to RHS columns

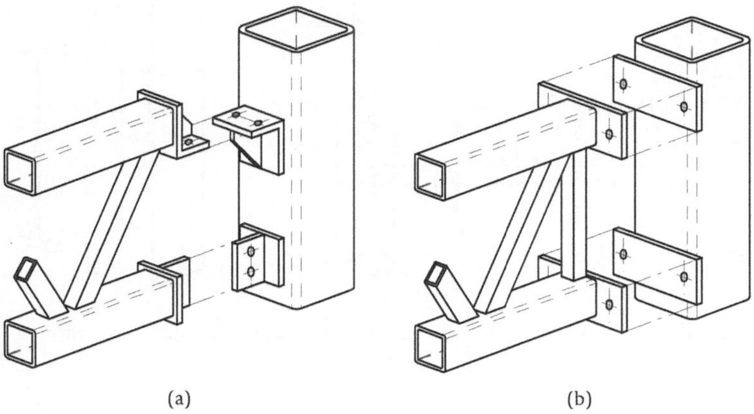

6.38 Truss to RHS column connections: (a) no end post; and (b) vertical end post

improve the local bending resistance of the wall of the incoming section.

For the connection of tubular trusses to RHS columns, typical bolted details are shown in Figure 6.38. High shear-forces may require the use of more bolts than shown. The sharing of load between the upper and lower chords in the connection depends on the presence of a vertical bracing member at the end of the truss. In the detail of Figure 6.38(a), the upper connection will resist all of the applied shear-force. In Figure 6.38(b), the upper and lower parts of the connections may be assumed to resist equal shear-force.

Whilst the above details may not be the most visually appropriate for exposed applications, they illustrate the general principles of support conditions to trusses using tubular sections of all types.

A good example of a simple and elegant detail of a connection between a CHS column and an I-section roof beam is shown in Figure 6.39.

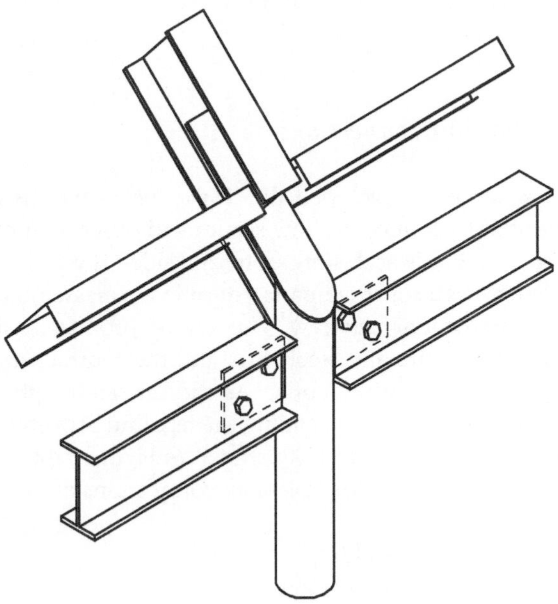

6.39 Details from Ball-Eastway House (architect: Glen Murcutt)

6.40 *Flowdrill* bolt

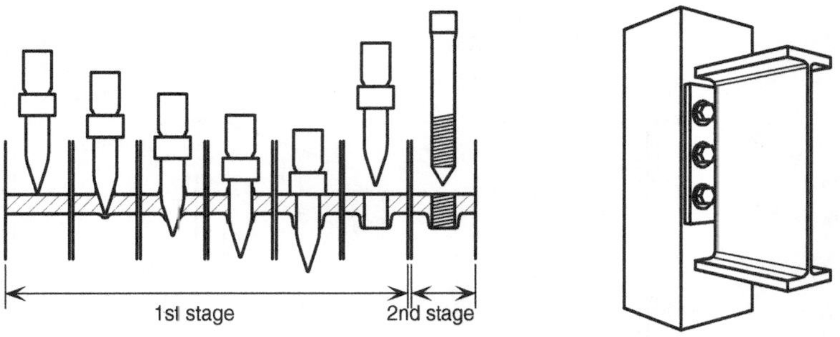

6.41 Illustration of the stages of forming and making a *Flowdrill* connection using a fully threaded bolt

6.10 Special bolted connections to SHS and RHS

6.10.1 Flowdrill *connections*

The '*Flowdrill*' method of bolting may be used where an architecturally 'clean' connection to an RHS member is required.

Flowdrill is a form of connection that does not require access from both sides of the connection. It is a thermal drilling process that makes a hole through the wall of a hollow section without the removal of the metal normally associated with drilling. The hole is then threaded in a second operation. The threaded hole will then accept a fully threaded bolt (see Figures 6.40 and 6.41). At present, the application of the *Flowdrill* process is limited to steel thicknesses up to 12.5 mm. It is mainly used for connecting end plates of beams to RHS connections.

Flowdrill requires the use of a high-speed drill, as the normal drill speed is not sufficient. Because of this, the RHS section may have to be taken out of the main production line, which adds both cost and time. Therefore, *Flowdrill* connections tend to be used for specialist applications. Further information can be obtained from Corus, Tubes & Pipes,[26] and guidance on connection design is given in *Joints in Simple Construction.*[5]

6.10.2 Hollo-Bolt connections

Lindapter has recently developed the '*Hollo-Bolt*', which is another type of bolt used to connect hollow sections to other members, and where the connection is accessible from one side only.

The *Hollo-Bolt* features three parts (supplied pre-assembled) — a body, cone and central setscrew. The entire product is inserted through both the fixture and steelwork, and the central set screw is tightened whilst gripping the collar. As the set screw tightens, the cone is drawn into the body, spreading the legs and forming a secure fixing. The *Hollo-Bolt* principle is illustrated in Figure 6.42.

The principal advantages of *Hollo-Bolt* connections are:

- there is no need for welding
- it is quick and simple to install

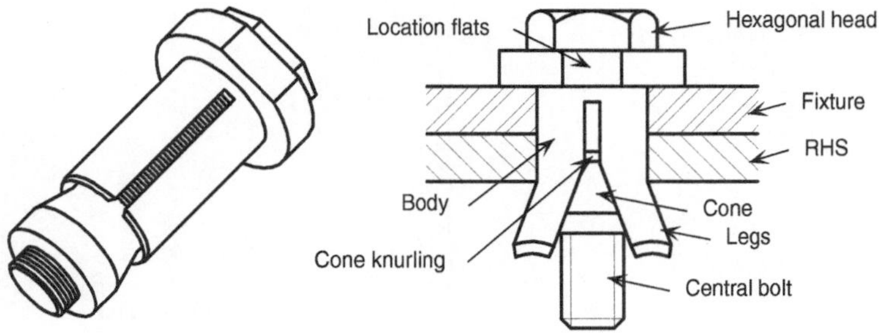

6.42 *Hollo-Bolt* connection

- it is fully tested in both tensile and shear applications
- no special tools are necessary, it can be installed using two spanners
- there is no need to provide close tolerance holes
- access is needed from one side only
- it is available in mild steel or stainless steel
- it can be used with a threaded rod or a central bolt
- no power is required on site.

The principal disadvantage of this connection is that the bolt hole is considerably larger than in normal bolted connections (approximately 1.7 times the bolt diameter), which may affect the local resistance of the wall of the RHS when subject to bending or tension forces. Furthermore, its capacity in shear and tension is low compared to normal grade 8.8 bolts.[5] Also, the bolt cannot be undone after it is tightened because of the expansion of the rear of the cone.

6.10.3 *Huck 'blind' fasteners*

The Huck Bom Blind fastening system uses fasteners between 3/16″ (4.8 mm) and 3/4″ (19 mm) diameter, which can be connected from one side only.

Resistances in shear and tension compare favourably with other kinds of connection, and the appearance is more attractive. The main disadvantage is that the fastener cannot be undone, and the connection appears to be more like a riveted than a bolted connection.

Chapter 7

Tension structures

Tension structures refer to suspended or 'tent-type' structures in which the 'ties', i.e. members designed to carry tension, are major elements in the overall structure. Tension structures differ from conventional framed-structures in two important respects: the structural concept is explicit in the architecture, and the detailing of the connection between the tension and compression elements can be more complex. The design of a tension structure requires careful thought about load paths, stability, flexibility of the system, cladding interfaces and foundation design.

In most framed buildings, the building itself defines the form of the structure to a large extent: columns, walls, beams and slabs are arranged and sized to suit the application using basic rules which are dictated by the plan form and structural efficiency. However, in a tension structure there is more freedom in the choice of the form of the structure, which is mostly external to the building envelope. Tension structures include various forms of suspended structures and cable-stayed roofs. Figures 2.9 to 2.11 and Figures 7.1 to 7.10 illustrate some well-known examples of these types of structure.

Tension structures are most commonly used in long-span roof structures, but they can be employed in a wide variety of applications, including canopies, glazed façades, and even staircases.

7.1 Tension-tie structure used to support a membrane roof at the Schlumberger Research Centre, Cambridge (architect: Michael Hopkins & Partners)

The principal advantages of tension structures are:

- they are a simple and efficient structural form
- their ability to create long-span enclosures
- they can be erected relatively easily
- their ability to accommodate flexible cladding materials or membranes
- they have discrete supports, leading to concentrated foundation forces.

Their disadvantages are partly related also to these advantages:

- heavy foundation forces both in compression (under the masts) and in tension (at the tie holding down points)
- additional space is required around the structure for the holding down arrangement
- the structural elements or ties often perforate the enclosure.

The supporting compression members in tension structures are commonly tubular sections. The attachment of the ties at the top of the masts is important structurally and architecturally. Often these compression members or masts are designed to resist multiple attachments to ties along their length.

Tension elements can be readily introduced into other forms of construction, which are not strictly 'tent-type' enclosures. These are:

- arch structures, with ties at their base or at intermediate locations
- portal frames, with ties at or close to eaves level (see Figure 3.12)
- the bottom chord of roof trusses, which is subject to tension (and also to compression due to wind uplift)
- tension elements in bracing systems.

7.1 Design opportunities for tension structures

The clear distinction between tension, compression and bending elements help to create structures with intrinsic visual interest. Tension structures can be very light visually since the sizes of the individual components are minimised.

The tension elements used in the primary structure are almost always external to the building and are 'expressed' visually. This is both from a desire to emphasise the structure, and also because it would be difficult to provide the depth or height required for the tension structure without an excessive increase in the building volume. Furthermore, an external structure reduces the visual bulk of the building by minimising the clad volume, and reduces the apparent bulk by breaking up the external surfaces. In auditoria, and sports stadia, tension structures lead to less visual obstruction for spectators.

Tension structures present a strong image that is appropriate for some projects and preferred by some clients, whilst at a small scale, the elements and details which form a tension structure provide interest and enrich the design. The aesthetics of the detailing of the tension system, and the principle of suspension and tension, can provide a motif which helps drive the whole design. The 1972 Munich Olympic Stadium by Frei Otto is a classic early example of this principle.

If a building is to be extended later, a largely external structure can be designed to reduce the disruption at the interface by making it possible to make the structural connections without perforating the enclosure.

Tension components are equally used in secondary structures, such as glass façades, where their lightness, delicacy and refinement of detail emphasise the transparency of the glass (see Chapter 9). Similarly, atrium roofs, staircases and footbridges are other examples in which a tension structure can be appropriate.

The reduction in component sizes can also help in transportation and erection. They can also be useful if heavy point loads are to be supported, as they may be connected to the suspension structure at designated points without penalising the whole structure. In the Fleetguard project, for example, the entrance bridge, stairs and roof-top mechanical plant are all suspended from the primary structure (Figure 7.2).

A tension structure may also provide a solution to problems related to particular site conditions. The Oxford Ice Rink is built on poor ground with a very low bearing capacity. The masts concentrated loads at two points and relieved loads elsewhere, which meant that expensive piled foundations were required in only two places (see Figure 7.3).

The tension structure of the Lord's Mound Stand is a response to the particular problem imposed by the site with the existing seating and roadway, and the need to minimise the obstruction to the spectators' view (see Colour Plate 25).

As noted earlier, a disadvantage of tension structures is often the need to provide heavy holding down positions at foundation level, and to protect the tension cables against vandalism, access, fire and corrosion. Penetrations through the external envelope should also be adequately weatherproofed (refer to Chapter 10).

7.2 Fleetguard, Quimper, in which heavy loads were supported by the primary structure (architect: Richard Rogers Partnership)

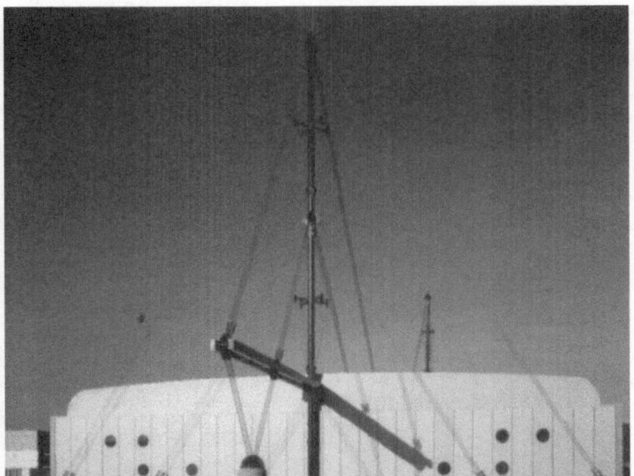

7.3 Oxford Ice Rink with discrete foundation supports (architect: Nicholas Grimshaw & Partners)

7.2 Different forms of tension attachments

In tension structures, the form and complexity of the attachments depends on the forces transferred, and the number, size and orientation of the individual ties. It is important to achieve a smooth transfer of tension force from the tie rods or cables into the supporting members, which may necessitate the use of relatively large connecting plates and other components, and which may dictate the detailing of the attachments.

The different 'generic' forms of tie attachments are:

- head detail at masts or columns by:
  - direct attachment
  - 'saddle' support
- intermediate attachment to masts or columns
- foundation attachments
- attachment to cantilevered beams
- intermediate tie attachments
- attachment of column bases
- cross-over ties.

Some general approaches to detailing of these forms of attachment are presented in Figures 7.4 to 7.9.

7.4 Forks at column head, Fleetguard, Quimper (architect: Richard Rogers Partnership)

7.5 Intermediate attachment to column in the Renault Parts Distribution Centre (architect: Foster and Partners)

7.6 Tension member attached to fabricated base, Hanover Trade Hall (architect: Thomas Herzog)

7.7 Tension anchors to cantilever beams at Sainsbury's supermarket, Camden, London (architect: Nicholas Grimshaw & Partners)

7.8 Fabricated fork detail at Igus Factory, Cologne (architect: Nicholas Grimshaw & Partners)

7.9 Column base attachment for tension rods at sports stadium, Cologne (architect: Verena Dietrich)

7.3 Fabric supported structures

Tension members are often used to support fabric membranes as tent-type enclosures. The structural solutions are designed to emphasise the 'lightness' of the form. The Schlumberger Research Centre, Cambridge (see Figure 7.1) and the Mound Stand at Lord's cricket ground (see Colour Plate 25) demonstrate the elegant use of tension structures to support flexible roof membranes. The same approach was used at a much larger scale in the Greenwich Millennium Dome (see Figure 1.3).

7.4 Adjustments

The assembly of the components in tension structures is relatively straightforward, and is a function of the discrete elements and their pinned joints. However, unlike conventional bolted structures, some form of adjustment must be provided in tensile structures to allow for assembly and movement, and to ensure that the enclosure is at the correct level and position. The adjustment system may also be used to control the distribution of forces within the system. It is wise to have as few adjustment points as possible so that the complexity of the operation is minimised.

7.5 Tie rod or cable connections

Two basic types of connection may be used in tension structures using tie rods or cables: those which connect tie rod or cables to each other, and those which connect to the main structural elements. Generally, both these forms of connection require a method of adjustment to prevent sag in the member and/or to induce a specified tension.

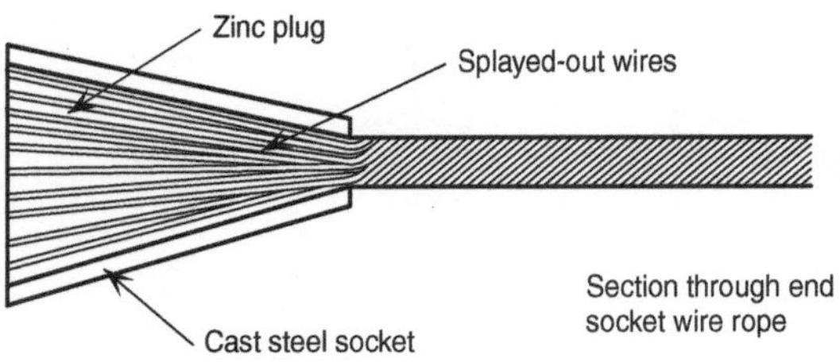

7.10 Splaying out of wire rope

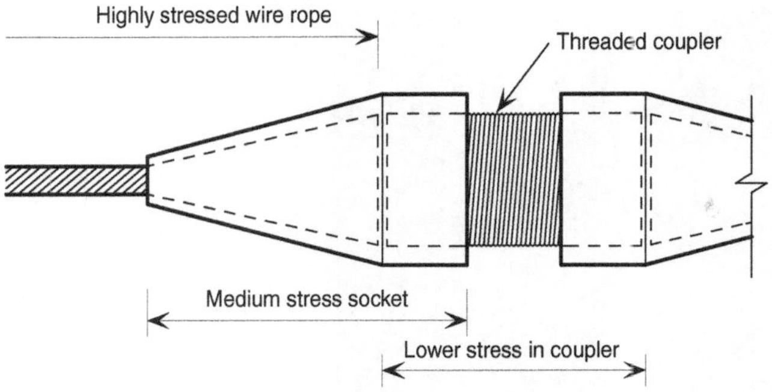

7.11 Threaded coupling for wire rope sockets

It is necessary, therefore, to provide an end detail to the cable or tie rod to which various fittings may be connected. This section reviews various forms of end details (terminations) and fittings which allow a range of connections to be made. Many of these connections use stainless steel components.

7.5.1 Cable terminations

Wire cables and ropes can resist very high-tensile forces but their ends cannot be threaded or welded. For low-tensile forces or for temporary connections, the cables can be clamped. For higher tensile forces, a number of methods are available, the most common of which are the socket termination and the swaged (or pressed) termination.

In a socket termination, the individual wires are spread into a conical-shaped steel casting (see Figure 7.10) and are anchored using zinc or resin. The casting can then be attached to any particular fitting or linking device by means of a threaded coupler (see Figure 7.11).

A swaged termination is usually provided by the manufacturer and details of various swaged terminations (with associated linkages) are reviewed in Section 7.5.6.

7.5.2 Tension bars

In tie bars or rods, the connections are formed by threading the bars. Welding is not appropriate as the bars are usually made from high-tensile steel. For the highest strength bars, the thread is not cut but is rolled onto the bar so that no cross-sectional area is lost. Left- and right-handed threads are formed onto the ends to enable couplings to be made, as shown in Figure 7.12. In tension rods, it is important

7.12 Threaded coupling for bars

7.13 Typical coupling between stainless steel bars

architecturally, as well as structurally, to be able to reduce any 'sag' in the rod, otherwise not only will the member not be fully effective in tension, but architecturally it will appear 'weak'.

The coupler is often provided with a flattened portion to assist in turning without damaging the surface. Figure 7.13 shows an example of a coupler in a large stainless steel bar used in the Helsinki Sanomat building. In this case, the thread has been cut into the bar.

7.5.3 Fork connections

A fork detail is an effective way of connecting two or more ties to single node points. Various forks are described in the following sections. It is generally easier to attach the fork to the rod or cable rather than the main member in order to simplify the fabrication.

The fork end can be simply connected via a pin onto a shaped steel-plate which is fixed to the supporting member or anchor block. The fork end is a pinned rather than a bolted connection, which permits quick and simple site assembly. The provision of a pin connection allows rotation and eliminates any induced bending stress in the tie. Some adjustment is provided within the fork end to allow for construction tolerance and to take up any sag. Fork ends can be supplied with left- or right-hand threads, as required. It is recommended that locking devices are used on fittings with opposite-handed threaded ends.

Forks are generally supplied fully threaded for maximum adjustment. Special caps can be provided to blank off the ends. Alternatively, the rod may be passed through and secured with a nut, but this results in a bulkier detail.

7.5.4 Pinned connection

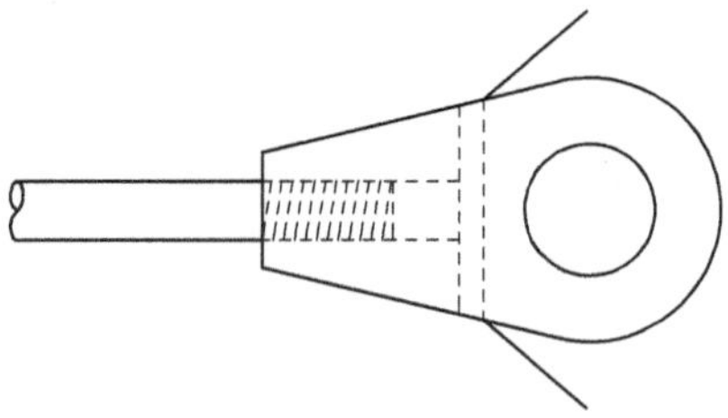

7.14 Pinned joint connection for tie bar

A conventional pinned detail of a bar to a plate is illustrated in Figure 7.14. These connections can be designed to emphasise the 'pin' form by using flatter coupling and larger 'pins'.

Corrosion protection of the connection may be achieved by over-painting the connection, by sealing it, or by using stainless steel pins. The latter only solves the problem of corrosion of the pin and neglects the plate and fork. There may also be a problem of bimetallic corrosion between steel and stainless steel, although this can be overcome by the use of isolating washers and bushes.

A good detail is to provide a recess at the junction between the fork and plate which can be pointed with an elastomeric sealant (see Figure 7.15). The ends of the pin should be sealed in the same way. This detail can easily be provided in both castings and steel plates.

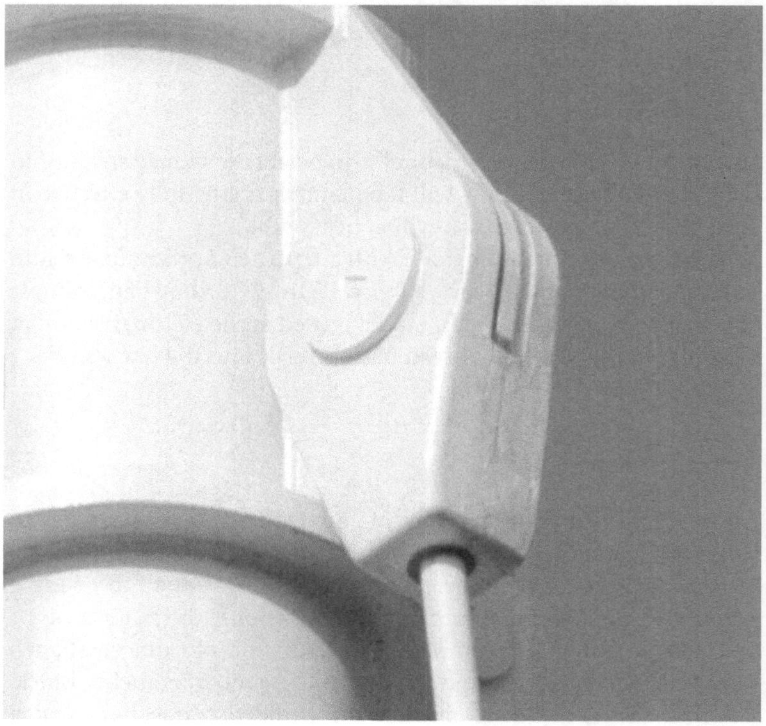

7.15 Sealant applied between fork and plate at the Renault Parts Distribution Centre, Swindon (architect: Foster and Partners)

7.5.5 *Other forms of end attachment*

Various forms of proprietary end attachments to cables and bars are reviewed below.

Spade ends

The function of the spade end is similar to that of the fork end. Spade ends can be used to connect cables or tendons into forks, or to directly connect to a plate fixed to the structure. They can also be used in the connections of twin cable ties, in order to minimise the bar size (see Figure 7.16).

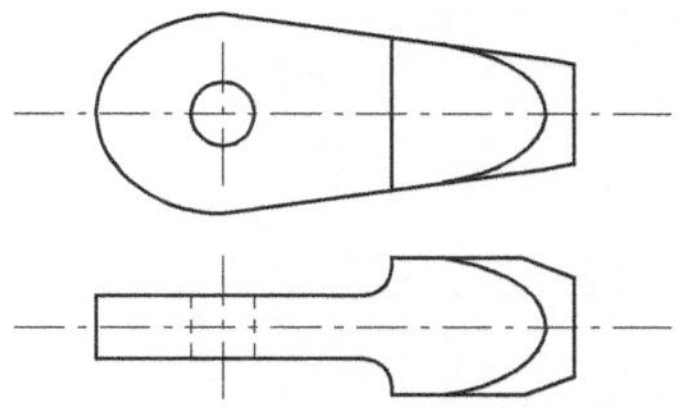

7.16 Spade end

Couplers

Couplers allow two bars to be simply joined together without any reduction of their load capacity. The end threads are both right-handed so that one bar may be connected to another in line. Couplers with left-hand and right-hand threads are available as specials (see Figure 7.12).

Turnbuckles

A turnbuckle provides a full strength connection between two rods with the additional provision of up to 50 mm adjustment. Turnbuckles are manufactured with left- and right-hand threads (see Figure 7.17), and are tightened whilst keeping the two rods in position.

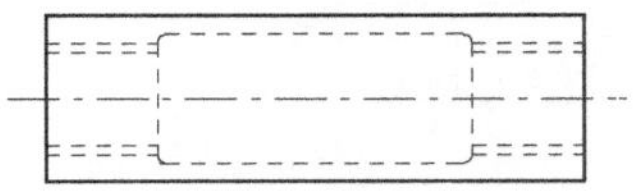

7.17 Turnbuckle

Pin sets

Pin sets allow the transfer of load from a fork or spade end to a node joint. A variety of pin types are available, which range from a high-quality fully machined pin and cover, through to a high-strength machined bar and circlips (see Figure 7.18).

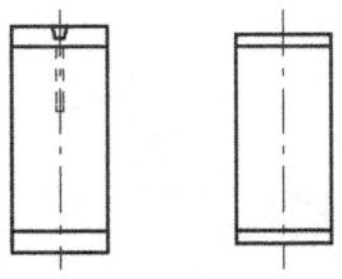

7.18 Pin set

Lock covers

Lock covers complement all the components listed above and provide an aesthetically pleasing locking device which also protects and hides exposed threads (see Figure 7.19).

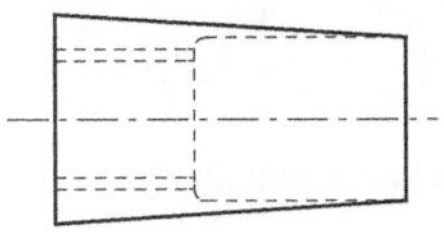

7.19 Lock cover

Specials

Special components and systems can be produced for specific projects in varying forms and to suit particular functions. Figure 7.20 shows a tie ring used to connect a number of ties and to transfer the tension forces amongst them.

For further reading on the design of tension structures refer to Chapter 20 'Tensile structures' of *Architecture and Construction in Steel.*[27]

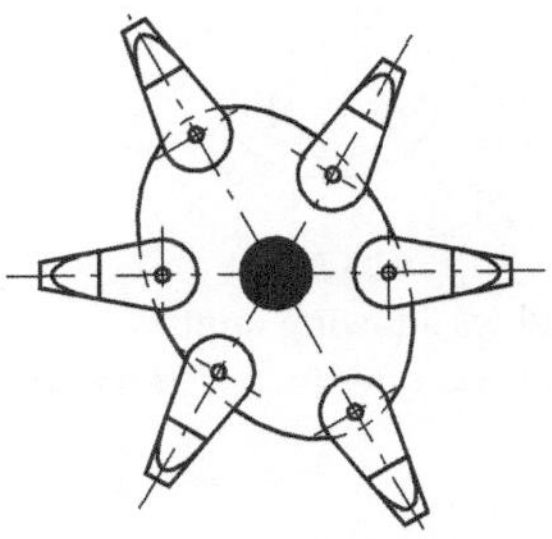

7.20 Bespoke tie-ring connection

7.5.6 *Proprietary bar and cable linking details*

A number of manufacturers produce cables with swaged (pressed) ends and with galvanized or stainless steel fittings. A typical

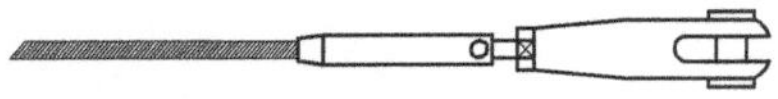

7.21 Swaged cable-end

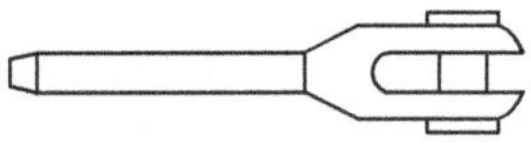

7.22 Fork end

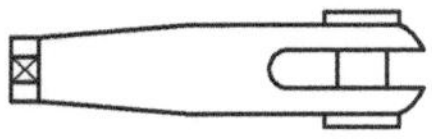

7.23 Adjustable fork-end

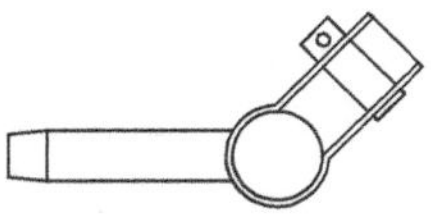

7.24 Toggle fork

7.25 Threaded end to permit adjustment

7.26 Swaged tensioner

7.27 Compact tensioner

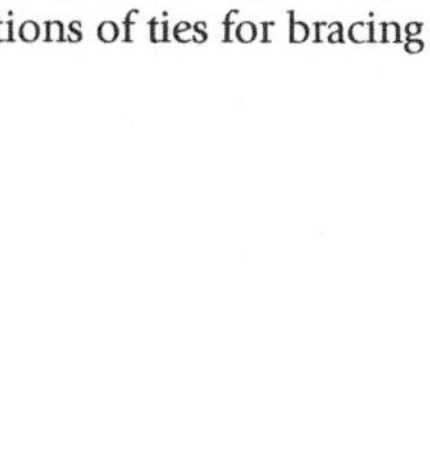

7.28 Rings allowing multiple connections of ties for bracing systems

swaged termination with an adjustable fork-end is illustrated in Figure 7.21.

Swaged fork-ends provide a slim and effective way of terminating a cable and bars, and provide a simple pin connection. The cable is anchored into the back of the fork using the swaging process, providing a neat and strong end. The fork end is simply connected by way of a clevis pin onto a profiled plate fixed to the structural member, or to an anchor block (see Figure 7.22).

Alternatively, threaded fork ends are used to terminate bar tendons. The bars are connected into the forks by a fine-machined screw thread.

Adjustable fork-ends have a fine-machined thread for attachment to bars. These forks allow adjustment for construction tolerances, or pre-loading of the tendon. The use of adjustable forks means that there is no need to use a separate tensioning device (see Figure 7.23).

Toggle forks can be used on cables, rods and bars, and provide effective connections with two planes of freedom, minimising any bending stress induced by the rotation of the tendon (see Figure 7.24).

Studs may be swaged onto cables and rods. The threaded end of the stud is screwed into matching spigots welded to the supporting member, or screwed into adjustable forks, allowing full adjustment or pre-loading (see Figure 7.25).

The standard single-point tensioner can be swaged to cables or rods. This achieves easier tensioning on site than adjustable forks, as adjustment or pre-loading is performed from a single point only (see Figure 7.26).

Single-point compact tensioners can be used on cables, rods or bars. They permit simple tensioning on site and are far more compact in length than the standard tensioner, but still allow similar adjustment (see Figure 7.27).

A variety of accessories is available which allows connection of multiple tendons. Rings can be supplied which allow the connection of pairs of ties for use as X-bracing. Complementing dome nuts are used to fix the ends of ties within the ring (see Figure 7.28).

Simple Delta plates may be used for connections of three tendons with fork-end fittings (see Figure 7.29). Similarly, rapid links allow quick and easy connection of single or multiple tendons, although they may seem relatively bulky in appearance (see Figure 7.30).

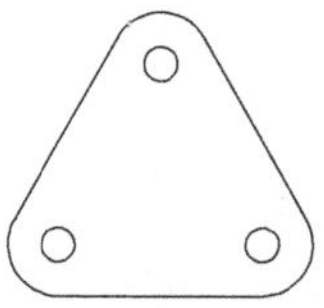

7.29 Delta plates for multiple connections of tendons with fork-end fittings

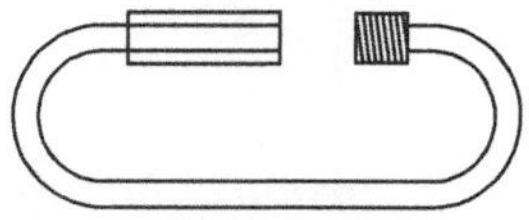

7.30 Rapid link

Stainless steel eye bolts are used as general-purpose fixing devices to masonry, reinforced concrete or steelwork (see Figure 7.31).

Isolation sleeves and washers should be used to prevent contact between dissimilar metals, thus reducing the possibility of bi-metallic corrosion.

Examples of the use of these cable-linking systems are presented in Figures 7.32 to 7.34. Figure 7.32 shows the use of a coupling device in an X-braced connection, also using a welded cruciform detail. Figure 7.33 shows a tension-coupling system for a stainless steel bar. Figure 7.34 shows the system used in the Millennium Dome for the attachment of the heavy duty tie-members to the foundations.

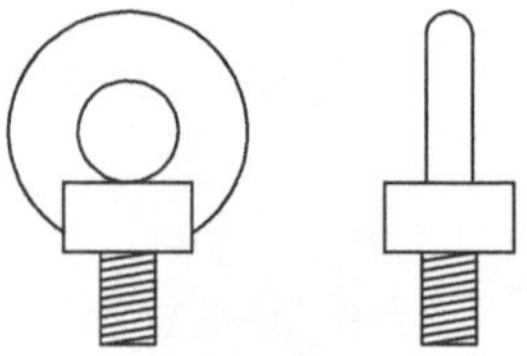

7.31 Eye bolt

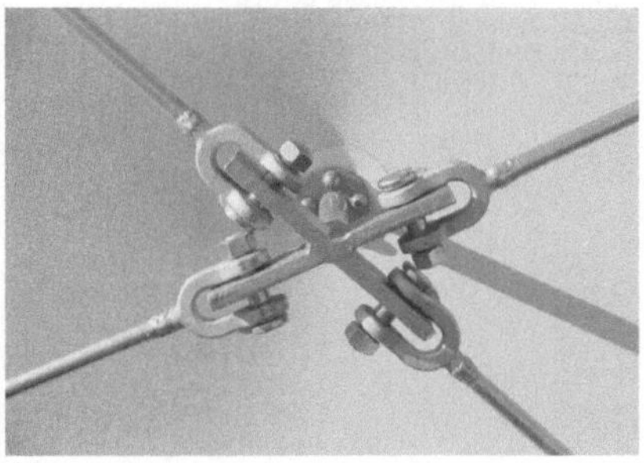

7.32 Coupler device used for an X-braced connection

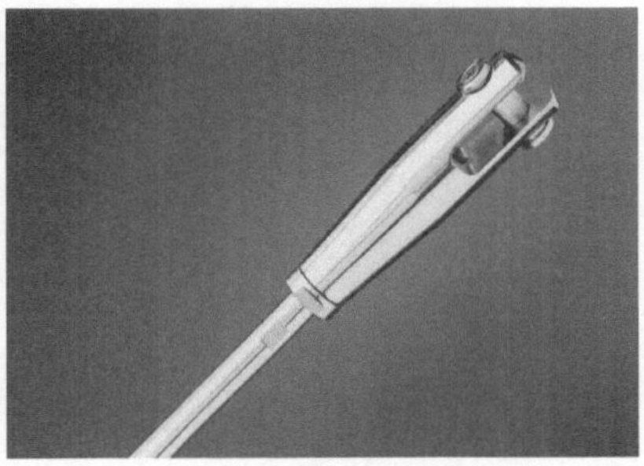

7.33 Tension-coupling system for stainless steel bar

7.34 Tension-coupling system for cables at the Millennium Dome, London (architect: Richard Rogers Partnership)

7.35 Multiple-cable connecting device

7.5.7 *Multiple-cable connections*

Multiple-cable connecting devices can be quite complex, as shown in Figure 7.35 which uses a cylindrical node. Another example of intermediate attachments between multiple cables is illustrated in Figure 7.36. At Chur Station, Switzerland, all the roof cables were connected at one complex node, as illustrated in Figure 7.37.

7.36 Intermediate attachments to columns, Schlumberger, Cambridge (architect: Michael Hopkins and Partners)

7.37 Complex node used at Chur Station, Switzerland (architect: Arup Associates)

7.6 Tension structures using tubular members

Tubular members are well adapted for use as the supporting or supported members in tension structures because:

- they possess good compressive resistance as masts, etc.
- they are more slender than other sections for the same load resistance
- the end details can be designed for cable attachments or saddles
- pinned or articulated end-details can be made, e.g. attachments to the foundations
- in some cases, the tubular members may also act as the tensile elements (particularly where load reversal may cause the same elements to be loaded in compression)
- tubular members possess good torsional properties and can resist torsional effects when curved on plan or in elevation
- cable attachments may be made along the length of the members (with suitable local reinforcement).

Tubular members are often used in external and 'expressed' structures. This is both from a desire to celebrate the structure, and because they provide the slenderness required without an excessive increase in the structural complexity.

Built in the early 1980s, the Inmos factory in Newport provided one of the first and well-known examples of the use of tubular steel trusses and masts in tension-tie structures. The 'wing-like' trusses are supported from the central spine of the building to create long-span flexible space internally. Figure 7.38 shows the form of the building

7.38 Inmos Factory, Newport (architect: Richard Rogers Partnership)

7.39 Tubular arch used to support a cable structure in the Hong Kong Aviary (architect: Hong Kong Architectural Office)

7.40 Local tie detail in Figure 7.39

7.41 Cantilevered arms of the Renault Parts Distribution Centre (architect: Foster and Partners)

and the multiple ties that were used. The Schlumberger building in Cambridge demonstrated the use of tubular masts to support the fabric roof, as shown in Figure 7.1.

Tubular structures can be designed to support cable-formed roofs by multiple attachment points. Inclined tubular arches are particularly successful, as indicated by the Hong Kong Aviary shown in Figure 7.39. The local detail of the attachment of the stainless steel cables is shown in Figure 7.40.

The cantilevered arms of the masts in the Renault Parts Distribution Centre are tapered and perforated to accentuate their lightness whilst serving to transfer compression forces to the tubular masts (Figure 7.41).

Chapter 8

Space frames

One of the most common forms of structural system employing smaller tubular sections is the 'space grid' or 'space frame', which are three-dimensional structural forms made using standard components and nodes.

The generic term 'space frame' is often used to describe two structural types: space trusses, with inclined 'web' elements, and space frames, comprising three-dimensional modular units. They both rely primarily upon full triangulation of the structure, provided the primary loads are applied directly at the node joints.

Most 'space grids' are based on the manufacture and assembly of standard elements and pre-formed node joint connections which can be easily transported and erected rapidly. Generally, tubular (normally circular hollow sections (CHSs)) sections are preferred for use in space frames because of their good compressive and local bending resistance. The elements are bolted together on site. A recent example of a space grid developed on a series of polygonal shapes is shown in Colour Plate 5.

8.1 Advantages and disadvantages of space grids

There are many benefits to be gained from the use of space grid structures, some of which are outlined below:

- loads are distributed more evenly to the supports
- deflections are reduced compared to two-dimensional structures of equivalent span, size and loading
- the open nature of the structure allows easy installation of mechanical and electrical services and air-handling ducts within the structural depth
- fixing details are greatly simplified — secondary members can be attached at the nodes and secondary elements such as purlins may not be required
- the structural indeterminacy of space grids means that, in general, failure of one element does not lead to overall failure of the structure
- modular space grids are usually factory made (thus producing accurate components), and are easily transportable and simple to assemble on site

- because of their modular nature, they may be extended without difficulty and even dismantled and re-erected elsewhere
- considerable freedom in space planning is achieved, although approximately square bays are preferable structurally, because they act like a two-dimensional grillage
- space frames can be assembled at ground level and lifted into place
- for ease of construction, most space grids have a regular grid pattern which may be exploited architecturally — particularly striking effects can be achieved if the colour chosen for the structure contrasts with the colour of the cladding.

Space grids are not appropriate for all roofing applications and their disadvantages may be summarised as follows:

- space grids can be more expensive than alternative structural systems, particularly when space grids are used for short spans (up to 20 m) or where there is no benefit of two-way spanning action
- the geometry is fixed, which can be a problem in irregularly shaped buildings
- visually, space grid structures appear very 'busy'; at some viewing angles the lightweight structure can appear to be very cluttered — grid size, depth and configuration can have considerable influence on the perceived density of the structure
- the number and complexity of the nodes can increase erection times on site — this is obviously very dependent on the system that is used and the grid module chosen
- when space grids are used to support floors, some form of fire protection may be required — this is more expensive to achieve economically because of the large number of relatively small-sized components.

8.2 Common forms of space grids

In space grid structures, where two plane grids are separated by inclined members, the top and bottom grids do not necessarily have to have the same pattern or orientation.

The common forms of double-layer grids are:

- square on square — where the top grid is directly above the bottom grid and the web members connect the layers in the plane of the grid lines (see Figure 8.1(a))
- square on square offset — where the bottom grid is offset by half a grid square relative to the upper grid, with web members connecting the intersection points on the top and bottom grids (see Figure 8.1(b))
- square on diagonal square — where the lower grid is set at 45° to the lines of support and is usually larger than the top grid and, again, with web members connecting the intersection points on the top and bottom grids (see Figure 8.1(c)). An

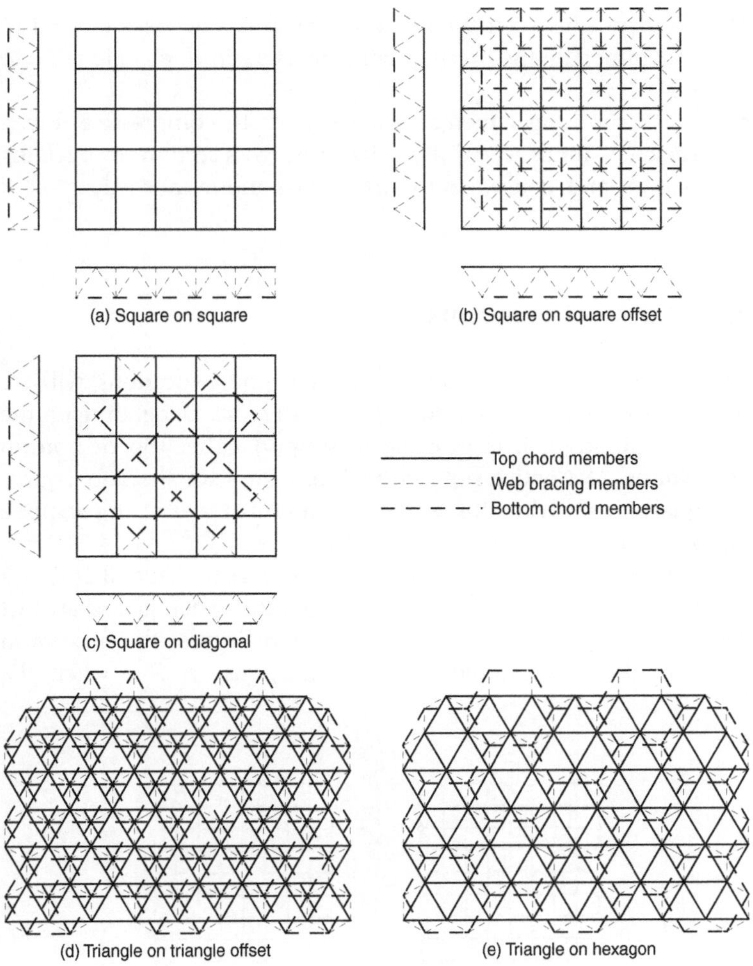

8.1 Different forms of double-layer grids

alternative version of this grid is diagonal on square where the upper grid is at 45° to the lines of support and the lower grid is parallel to the supports

- triangle on triangle offset — where both grids are triangular but the lower grid intersections occur below the centroids of alternate triangles in the upper grid, with web members connecting the intersection points on the top and bottom grids (see Figure 8.1(d))
- triangle on hexagon — where the upper grid is triangular and the lower, more open, grid is hexagonal due to the removal of some joints and web elements from the grid type described in the bullet point above (see Figure 8.1(e)).

The choice of grid configuration and depth will affect the economy of the space grid, as the node joints are usually the most expensive components. Therefore, the more node points in a given plan area, the higher the cost is likely to be. Increasing the grid-module size reduces the number of nodes for a given plan area, but there may be adverse consequences, as follows:

- the depth between the two grids may have to be increased to accommodate the bracing members at a sensible angle of 30 to 40°
- when the longer members are subject to compressive forces, they will almost certainly be larger in cross-section for buckling reasons and, consequently, are heavier and more costly.

8.3 Support locations

The choice of the most advantageous support locations will, of course, depend on the plan form of the structure. Depending on the grid configuration, it is possible to support either top or bottom node joints. Alternative support positions for a square plan, square on square offset-grid roof structure, supported at the upper node joints, are shown in Figure 8.2(a)–(d).

The provision of continuous edge support, as in Figure 8.2(a), is a more economical method of support for the space grid than just corner supports, as in Figure 8.2(b). This is because the maximum forces in the space frame and its deflections are less when the

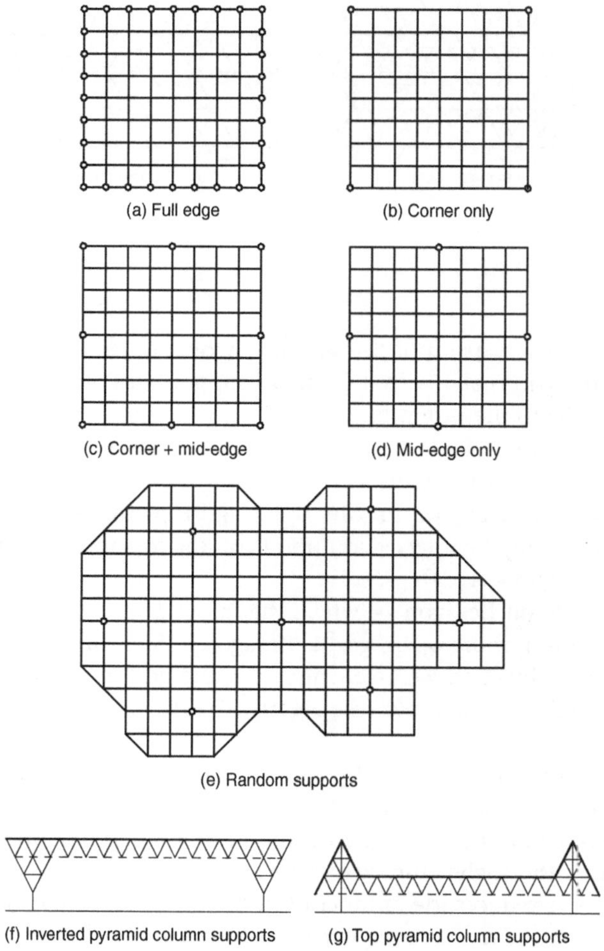

8.2 Typical support locations for space frames

supports are continuous. Intermediate supports along each edge, as in Figure 8.2(d), may also produce an efficient support system. In this case, some of the bottom-layer members will be in compression and some of the top layer will be in tension. To reduce deflections still further, the supports can be brought in slightly from the edges of the space frame to produce a cantilever portion around the whole structure. This can be useful architecturally as it allows the opportunity to have apparently column-free elevations.

An alternative method is to use 'tree' supports instead of individual columns. Commonly, this is achieved by inverting a grid module (e.g. a pyramid) at each support location, as the grid is then supported on several nodes at each column location, which reduces the forces in the bracing members at the supports. Irregular plan-forms can be created using modern analysis techniques, although the fabrication costs will be higher if more irregular member lengths and sizes are required.

8.4 Span:depth ratios

It is difficult to generalise about overall span:depth ratios for space grid structures as they depend on the method of support, type of loading and, to a large extent, on the system being considered. Makowski[28] suggests that span:depth ratios should be reduced to approximately 20 when the supports are only at, or near, the corners of the grid. Span tables produced by Space Deck Ltd indicate that for typical loadings, roof span:depth ratios of about 30 are possible using standard modules when square on plan. For example, a square roof supported on all four sides can span up to 40 m^2 based on a module that is only 1200 mm deep.

8.5 Commercially available systems

Although all types of space frame are not covered in this publication, the main systems currently available in the UK are described below.

8.5.1 Nodus

The Nodus system (originally developed by Corus and now owned by Space Decks), uses cast steel connectors for the Nodus joints, which are butt-welded to the chord or bracing members in jigs to ensure dimensional accuracy. Chord members are clamped together between the halves of the node casings using high-strength friction grip bolts, and the bracing members are joined to lugs on the node plate by steel pins through the forked-end connectors. The node configuration, and the pinned connections of the bracing members, permits variation of the depth of the space frame. The nodes are only produced in two forms. Lugs for the connection of bracing members are either in line with the chord members or, alternatively, at 45° to

8.3 Canopy at Terminal 2, Manchester Airport

the chords when viewed in plan. This limits the possible grid configurations to variations of the square on square, or square on diagonal layouts.

An example of the use of Nodus is in the roof of Terminal 2 at Manchester Airport (see Figure 8.3).

8.5.2 Space decks

This system originally developed in Australia consists of pyramidal units constructed from a square frame of steel angles connected by circular steel tube bracing members to a cast steel boss. All elements of the pyramids are welded together in a jig to ensure dimensional accuracy. The boss at the base of the pyramids has two threaded holes in one horizontal direction and two threaded studs in the other to receive connecting steel tie-bars.

Assembly is achieved by bolting the angle frames of the pyramids together and then connecting the cast steel bosses with tie bars. The tie bars have left-handed threads at one end and right-handed

8.4 Bentalls Centre, Kingston-upon-Thames (architect: BDP)

threads at the other, as do the holes and studs on each side of the cast bosses. Rotation of the tie-bar screws the bar into the boss at each end simultaneously. By varying the length of the tie bars, the node spacing can be altered to produce a one- or two-way camber in the space frame. An example of this form of construction is at the Bentalls Centre, Kingston-upon-Thames (see Figure 8.4).

8.5.3 MERO

The MERO system is an elegant concept in which the circular tube members connect to 'ball' joints at the nodes by a single concealed bolt. The name MERO derives from an abbreviation of the original name Mengeringhausen Rohrbauweise.

The circular steel sections have tapered cone sections welded to each end (complete with connection bolt and sleeve), and the nodes are hot-forged solid steel or drop-forged aluminium spheres with drilled and tapped holes which are profiled to receive the tube ends. The detail of a MERO KK ball is shown in Figure 8.5.

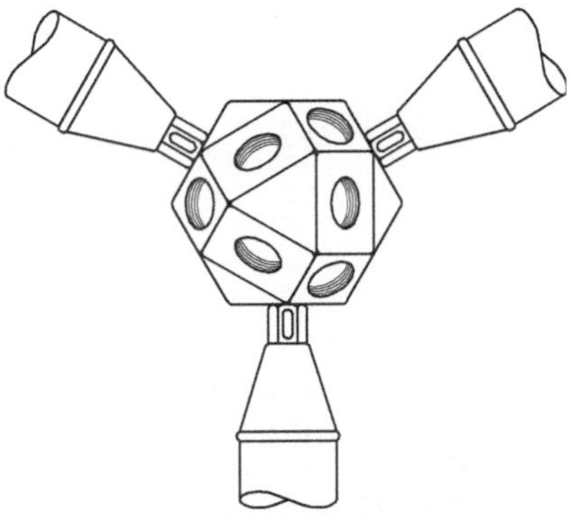

8.5 MERO KK ball

The National Indoor Arena for Sport at Birmingham has a triple-layer MERO space truss roof of 90 m × 128 m plan, and varies in depth from 8 m to 10 m.

The MERO system was also used for the Eden project in Cornwall, in which hexagonal 'cells' were created and joined together to form a large spherical surface (Figure 8.6).

A recently completed theatre complex in Singapore created a shell-like structure where solar shading was included in the space frame networks around the periphery of the building (Figure 8.7).

In miniature, the MERO system is also used for shop displays and exhibition stands.

8.6 MERO space-frame node used in the Eden Project, Cornwall (architect: Nicholas Grimshaw & Partners)

8.7 Mero space frame incorporating solar shading ,used in the Esplanade Theatre Complex, Singapore (architect: Michael Wilford & Partners and DP Architects Pte Ltd)

8.5.4 SPACEgrid

SPACEgrid is a space truss modular system which was developed from the UNIBAT system. To refer to SPACEgrid as a system is perhaps a misnomer as there is no standard module or joint. The most economical grid is selected according to the plan dimensions and loading, and the joints are designed to suit the grid layout and member sections used.

8.5.5 CUBIC space frame

The CUBIC space frame is a modular system in which an orthogonal grid is formed from Vierendeel girders in both directions, and joints are introduced in the top and bottom chords midway between each chord intersection. The grid can be constructed of modules 'X', 'T' or 'L' shaped in plan. These are the basic modules of the CUBIC space frame system which uses both tubular and open sections.

In 1988, the roof of a maintenance hangar for Boeing 747 aircraft at Stansted Airport was constructed from the system. The modules were 4 m deep and the space frame covered a diamond shaped plan of 98 m × 170 m on the major axes, without internal supports. (The CUBIC space frame is no longer available.)

8.5.6 Arched triangular lattice grid

One system which, up to now, has remained at the concept stage is an arched triangular lattice grid proposed in the competition winning design for the athletic stadium in Frankfurt, by Foster and Partners (see Figure 8.8). The arch roof shape was chosen and the supporting framework was conceived as a triangular plan grid that offset the longer route of arching forces with the provision of more frequent-braced points. It was manufactured as a series of identical prefabricated units in the form of individual 'diamond' shapes.

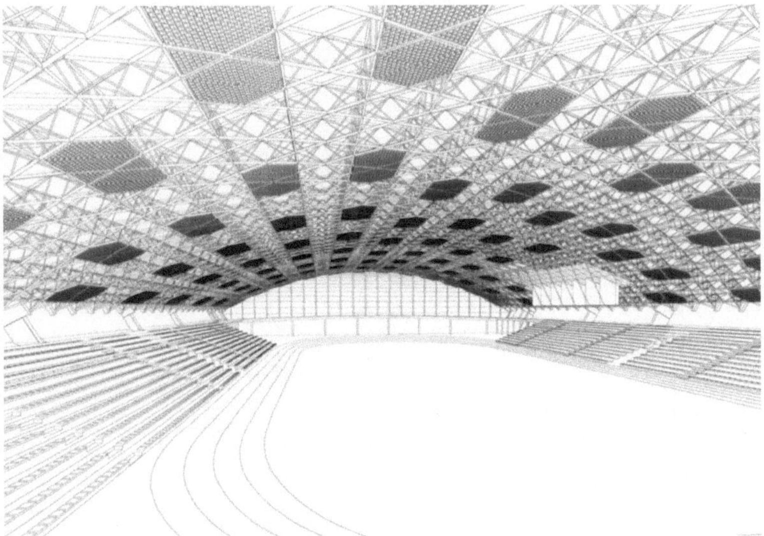

8.8 Competition winning stadium design, Frankfurt (Architect: Foster and Partners)

Chapter 9

Glazing interface details

9.1 Architecture

Since the technical and commercial development of large-scale glazing systems during the second half of the nineteenth century, the notion of transparency has exerted a seductive hold on the architectural imagination. Since then, the increasing sophistication of glass has presented architects with new and richer possibilities.

The positioning of windows and glazing is not just a response to the practical issues of providing light, external and internal views, and, where possible, natural ventilation. The therapeutic quality gained by the capturing and framing of light are often the essential ingredient which can transform an otherwise banal or sombre space.

With a thorough understanding of building physics, climatic influences, location and control of glare, the benefits of large areas of glazing can be exploited without excessive solar gain during the daytime or heat loss at night. Often perceived as weightless, the attributes of a glazed or translucent fabric roof can bring the interior closer to the sky and enhance an inter-relationship with nature. Modern atria within and between buildings use a wide range of glass-supporting structures, including curved and inclined forms (see Figure 9.1).

The introduction of glass also represents a contrast from the solid elements of a building. Thus, the quality of an internal and external space is achieved by the manipulation of the key components which can exert a powerful influence on the quality of the space, and the way it is viewed externally, as in Figure 9.2.

Steel can provide the essential support framework in a minimal way that does not detract from the appearance or function of the glazing. The pursuit of more advanced glazing and solar control systems has spawned the increased use of specialist steel products, and has advanced the development of intelligent blinds, louvres, grilles and panels.

At the Western Morning News Headquarters in Plymouth, the perimeter transparent planar glass wall is achieved by the delicate connection of the panes of glass to steel tusks via cast connecting arms, interlinked by rod rigging (see Figure 9.3).

9.1 Glazing attachments to curved roof beam (architect: Jestico and Whiles)

9.2 Shell structure clad in glazing at the Museum of Fruit, Yamanashi, Japan (architect: Itsuko Hasegawa)

9.3 Sloping glazing details at Western Morning News, Plymouth (architect: Nicholas Grimshaw & Partners)

9.2 Interfaces

The interfacing of glazing with other elements of construction is an important consideration, as each material component will have its own design, procurement and dimensional attributes. Guidance on interfaces is given in the SCI publication *Steel Supported Glazing Systems*[29] and in *Structural Use of Glass*.[30] The direct attachment of the glass panel to the support structure is often at discrete points rather than being continuous along the edges in more conventional glazed panels.

Pilkington's '*planar glazing*' system uses special bolted attachments which provide a more uniform bearing pressure on the glass, thereby increasing the local resistance to in-plane and out-of-plane forces. These attachments are not generally directly supported by the primary structure, but indirectly through a secondary element which provides for some articulation and adjustment for tolerances.

Planar glazing is often combined with articulated nodes which attach two or four panels of glass at their corners. The planar glazing system used at Heathrow Airport Visitor's Centre was attached to the supporting steel sections by means of stainless steel fixings, and provided a fully glazed yet acoustically sealed corridor along the runway side of the building (see Figure 9.4).

9.4 Support to glass wall at the Heathrow Airport Visitor's Centre (architect: Bennetts Associates)

The detailing of the interfaces between the glazing and its support structure is critical and should take into account:

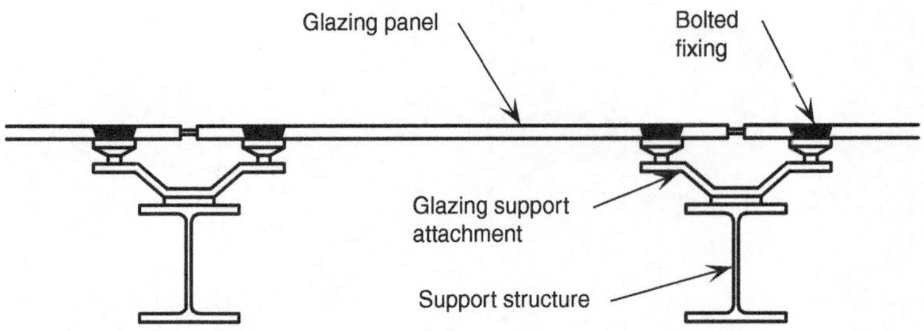

9.5 Typical bolted attachment to glazing showing the various components

- appropriate construction and manufacturing tolerances
- the movement of glazing panels and support structure due to thermal effects and wind or other applied loadings
- the loads that arise due to the self weight of the system and external actions — the load transfer effects that may occur when glazing panels are broken or removed should also be considered.

If movement cannot be accommodated, stresses are developed that can result in the distortion of members or, more likely, the breakage of the glass panes. The local attachment details should provide for the necessary adjustment. The detail in Figure 9.5 achieves this by use of slotted holes and adjustable bolted attachments to the supporting members.

9.3 Tolerances

Although glazing support systems are usually specialist elements, the interface with the steel supporting members is critical in both structural and architectural terms.[29,30] Furthermore, stricter tolerances may be required for the primary structure in order to satisfy the requirements of an elegant detail.

Glazing panels can be made to a high degree of accuracy, to between ±2 mm for single glazing and ±4.5 mm for double glazing, whereas the supporting steelwork may at best be constructed to ±5 mm tolerance on the overall length.

As the lead-in time for some specialist glazing systems is around 12 weeks, there is little possibility of the glazing sub-contractor being able to take site measurements, and the tolerances and erection sequences have to be addressed at the design stage. The connection detail must therefore allow for differences in the accuracy of the erected components.

Essentially, allowance for adjustment within normal construction tolerances, and for later movement and deflection, can be accommodated using slotted connections. These can be designed in such a way to allow for both horizontal and vertical movement. The actual provision for movement around the bolt-holes will depend upon a number of factors, such as glass thickness, number of bolts and the size of the glazing panel. Specialist suppliers can advise on the permitted limits of deflection, depending upon the size and shape of glass panes as well as the overall area of glazing.

9.4 Support structures

Various examples of support systems to glazed walls are illustrated in Figure 9.6. They can be fabricated from a variety of I-section shapes, but tubular members are generally preferred.

The choice of system will depend on aesthetic considerations, as well as cost and structural requirements. In a large expanse of glazing, architects may prefer to minimise the bulk of support structure to enhance its degree of transparency. The supporting steelwork may be combined with the use of rods or cables for tension members, which may be spaced from the compression members by struts. In a trussed system, the tensioned cables provide resistance to bending by their 'push-pull' action with the compression members (which is often a tubular section).

9.4.1 *Trusses*

Trusses are one of the most commonly used support systems, particularly in long-span applications. They are termed 'wind trusses' because their principal action is to resist wind loads. The generic forms of wind trusses are illustrated in Figure 9.7. They may span vertically and/or horizontally, and are generally installed at spacings corresponding to the width of the glazing panels. The spacing of the wind trusses may be increased by introducing cantilevered arms or other secondary members but, in this case, the flexibility of the arms will increase the movement of the glazing.

Lateral restraint may often be provided by tension systems using wires or rods. Turnbuckles or similar devices can be incorporated to provide length adjustment, and can be used to pull the trusses or fins into correct alignment (see below).

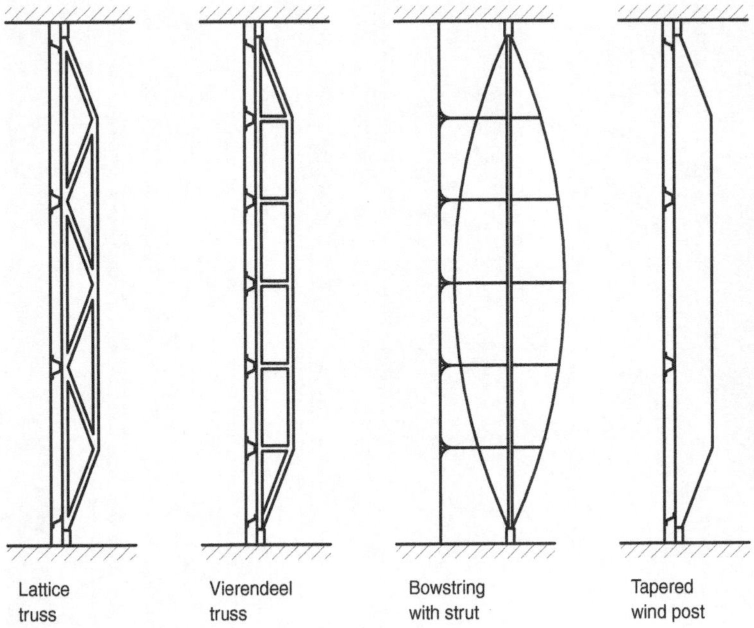

9.6 Types of support systems for large glazing panels

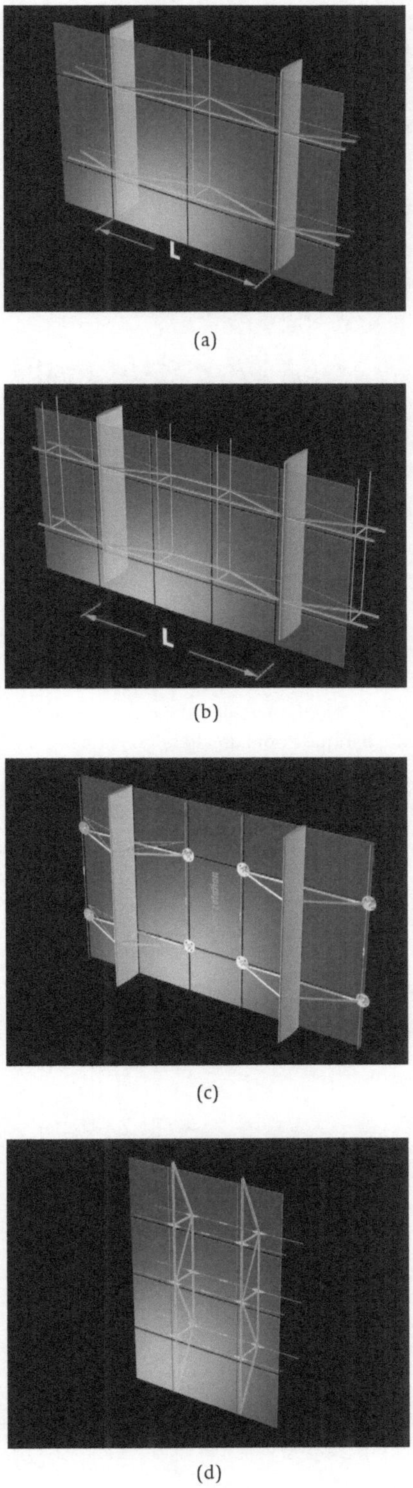

9.7 Wind-truss support systems: (a) horizontal wind truss; (b) longer span wind truss; (c) cantilever supports; and (d) vertical truss

9.4.2 Tension systems

Support systems can be constructed almost entirely from tension elements, where required, such as rods or wires. Linking systems have inherent adjustment to allow for construction tolerances. Pinned connections allow the ties to be quickly and simply installed with the added benefit of allowing rotational movement to minimise any induced bending of the ties. A variety of components, often in stainless steel, are used in rod rigging and cable trusses, as shown in Figure 9.8. Examples of tension structures used as secondary elements include the glass walls at the Banque Populaire in Rennes, shown in Figure 9.9, and at the University of Bremen, shown in Figure 9.10.

However, the boundary supporting structure for tension systems can be heavier than in other systems, as the high-tension loads have to be transferred to stiff supports at both ends of the cables or rods. The boundary structure can be in the form of a horizontal or inclined steel truss or, alternatively, a direct connection to a foundation.

Cable systems may also be used with other forms of attachment. The covered courtyard of the History Museum in Hamburg was conceived as a lightweight structure consisting of cables and flat-plate connections, which directly supported the glazing panels, as shown in Figures 9.11 and 9.12.

9.8 Tension-coupling system used to support a fully glazed wall

9.9 Double skin glazed façade used at the Banque Populaire, Rennes, France (architect: Odile Decq and Benoit Cornete)

9.10 Tension structure to support glazing at the University of Bremen, Germany (architect: Jan Stormer)

9.11 Courtyard at Hamburg City History Museum (architect: Von Gerkan Marg & Partners)

9.12 Detail of glazing in Figure 9.11

9.4.3 *Support attachments*

Although glazing can be attached directly to the support structure, it is more common to use separate components, called glazing support attachments, that form an interface to allow easier adjustment in position, and which are less reliant on the accuracy of setting out of the main structure. Glazing support attachments are normally located at the node points of trusses. Glazing panels can be attached to one side of lattice and Vierendeel trusses, or occasionally on both sides where a twin-skin façade is required.

Glazing panels supported on bowstring trusses may either be positioned to one side of the truss (in which case they are often attached to extended struts, as in Figure 9.6) or set within the truss (in which case the system should be detailed to allow the truss members to pass through the plane of the glazing).

Glazing support attachments may take many forms, but the most common are:

- angle brackets
- spiders
- pin brackets
- clamping devices.

9.13 Glazing support attachments:
(a) attachment to tubular sections; and
(b) 'Spider' attachments

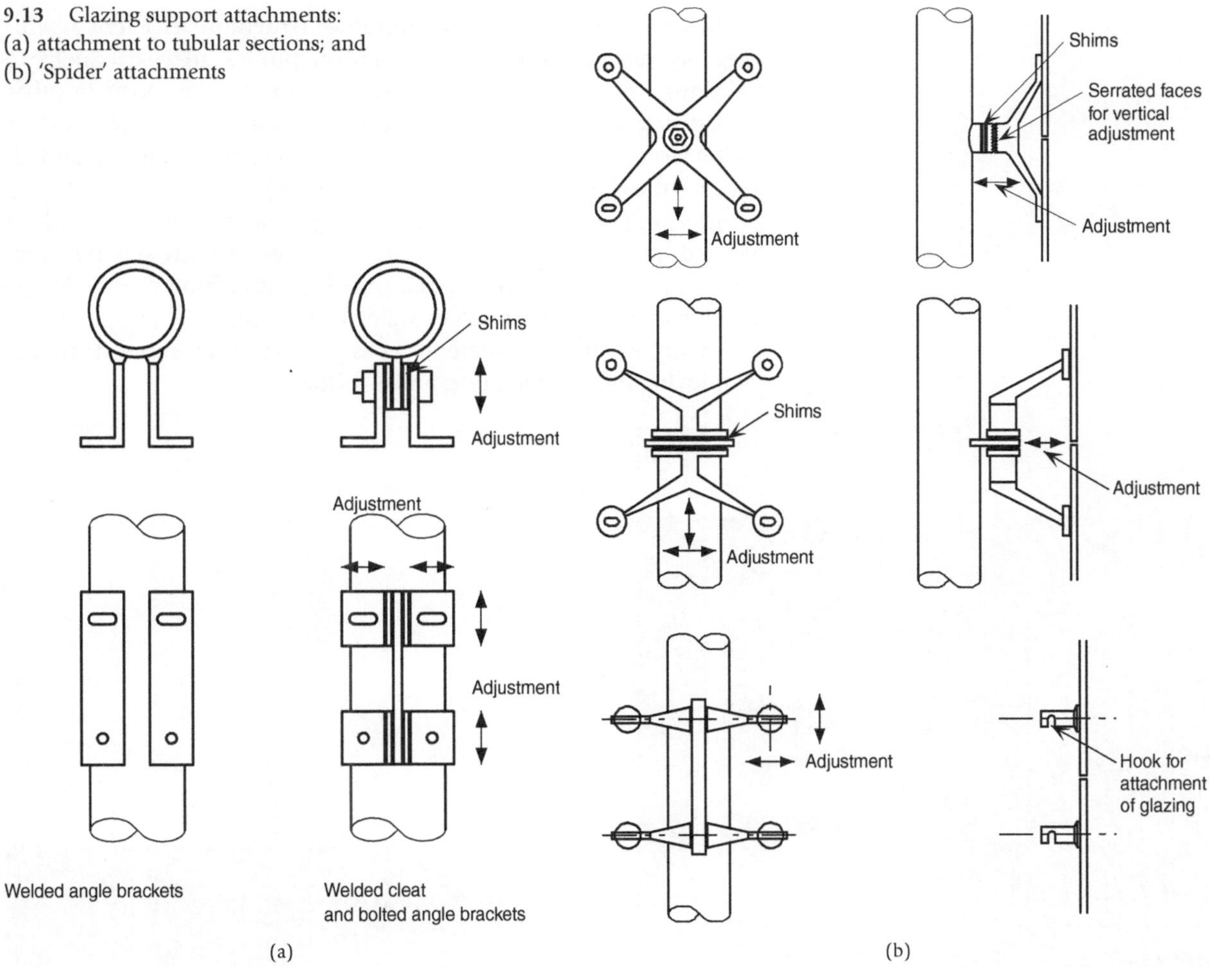

These alternative devices are illustrated in Figure 9.13. Simple attachments to tubular connections, as in Figure 9.13(a), are relatively stiff, whereas 'spider' attachments, as in Figure 9.13(b), are more flexible.

Castings are often used for the connection nodes in 'planar glazing' systems. These castings can be steel or aluminium, although stainless steel is often preferred.[31] Standard castings are readily available from some suppliers, but bespoke fittings can add significantly to the lead-in times for these components.

9.5 Use of tubular members in glazing systems

Tubular sections are often used to provide support to large glazed panels in walls and roofs. Guidance is given in the SCI publication *Steel Supported Glazing Systems*.[29] Specific examples related to the use of tubes are included in this section.

Tubular steel members combine structural efficiency with elegance and, because of their section profile, they also help to minimise the apparent bulk of the structure. Trusses, posts or other supporting steelwork may be combined with the use of rods or cables as tension members to enhance the architectural appeal, and to create striking visual effects (see Figure 9.14).

The support attachments for the glazing are often bolted or welded attachments to the tubular members, or screwed into pre-formed bolt holes in the tubular members. Attachments can also be clamped rather than bolted or welded to the tubes.

Examples of the different types of structural support system employing tubular members are as follows:

9.14 Glazing to Greenock sports centre (architect: FaulknerBrowns)

9.15 Tubular mullion used to support a tall glazed wall (architect: RFR)

9.16 Curved truss used to support the roof of the Department for Trade and Industry building (architect: DEGW)

9.17 Inclined trusses to support a glazed wall

- mullions to support glazed walls — the example in Figure 9.15 shows the use of extended arms to support the tall glazed wall
- compression chords of cable-tensioned glazing systems, as in Figure 9.6
- roof trusses of various cross-sectional forms (see Figure 9.16)
- trusses used as vertical or inclined mullions (see Figure 9.17)
- curved members to support separate glazing support arms (see Figure 9.18)
- curved trusses, with 'faceted' glazing systems, as in Waterloo International, shown in Figure 9.19
- tubular lattice with direct attachment at the node points.

9.18 Curved glazing support arm used at Frankfurt Airport's IC station

9.19 Glazing detail at Waterloo International station (architect: Nicholas Grimshaw & Partners)

In the Leipzig Trade Fair, shown in Colour Plate 6, the glass panels are suspended from the main steel tubular arches by cast steel arms connected to stainless steel fingers. Adjustable screws at the connection accommodate out-of-plane tolerances, whilst adjusting plates accommodate in-plane tolerances.

At Waterloo International station, the glazing panels are supported on stainless steel rods which are attached by brackets directly to the tubular steel members (see Figure 9.19).

A filigree of steel cables and small plates can produce a very delicate spider's web-like structure ideal for a glazed roof (see Colour Plate 11).

Chapter 10

Steelwork penetrations of the external envelope

An exposed structure is devised either as a completely independent structure supporting the external skin and other elements, such as canopies, or as an extension of an internal primary structure. In the second case, there are instances when it is necessary to penetrate the cladding to make connections between the internal and external members. Typical details are reviewed in the following section.

10.1 Waterproofing

A suspended roof almost inevitably necessitates penetration of the waterproofing layer of the building fabric. However, a typical industrial building with a conventional internal structure has a number of penetrations for rainwater outlets, smoke vents, flues, etc. Thus, it is a change in the 'size' of the penetration, rather than the nature of a new problem. In practice, it is found that leaks occur most often at rainwater outlets and parapets. In contrast, it is relatively straightforward to design a waterproofing detail for a column or hanger penetration.

In order to make the detailing as simple as possible, the primary structural connections of an external structure should be located outside the building envelope, and penetrations minimised. Where tie rods pass through the roof, it is difficult to dress the waterproofing around such small members. In this case, it is often necessary to provide a shroud for the waterproofing to tuck under, as shown in Figure 10.1.

10.1 Waterproof shroud used in roofing

10.2 Cold bridging

Some designers may be concerned that steel penetrations through the building envelope may form a 'cold bridge' which can lead to condensation, or in extreme cases, to a risk of corrosion. There is no answer that is appropriate for all applications and the designer must assess the risks and consequences of the cold bridging. If the internal

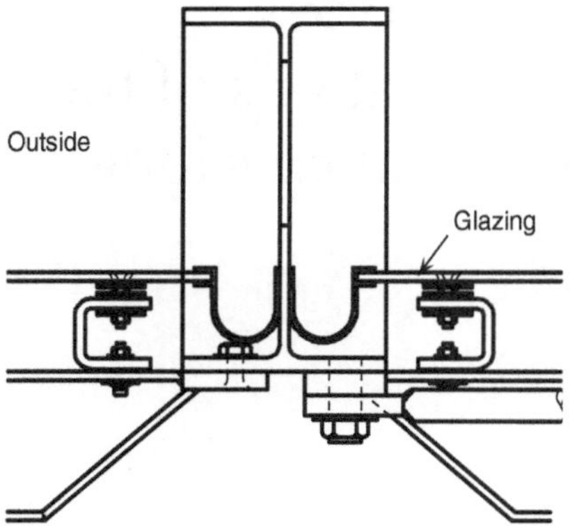

10.2 Steel column extends outside the perimeter wall, with the interior weather protected by rubber gaskets — Renault building, Swindon (architect: Foster and Partners)

atmosphere is not very humid, and the exposed steel member is not adjacent to materials likely to be damaged by moisture, then experience has shown that steel members passing through the envelope should not lead to problems of cold bridging.

If there is a major concern, an element of trace heating can be introduced along the steel member just inside of the external skin, although this is rarely used in practice. Alternatively, the steel member can be joined at the point of the cladding to another section with an insulating component placed between the two pieces. However, this can be very difficult to resolve elegantly if both inside and outside components are exposed to full view. The following examples illustrate details that have been employed successfully.

Figure 10.2 shows a section through an external wall of the Renault Parts Distribution Centre, which is illustrated in larger scale in Figure 1.2. The perforated vertical members support both the cladding and glazing. At the ends of the glass panels, rubber gaskets abut the steel section and act as a weatherseal.

At the David Mellor Cutlery Factory, a steel bracket was inserted in the plane of the glass. The bracket connects the internal roof structure with a solid steel perimeter ring bar, as shown in Figure 10.3.

Studio Downie's Visitor's Centre in Sussex uses the interplay of planes created by solid and transparent walls to heighten awareness of what is outside. The modest sized steel-frame penetrates the glass skin, which softens the edges between inside and out, as shown in Figure 10.4.

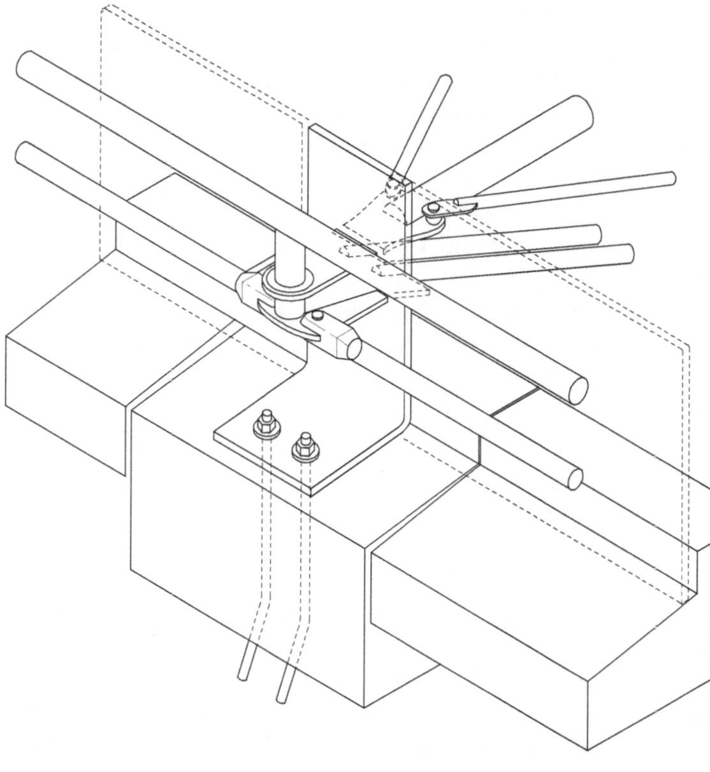

10.3 Detail of steel bracket connecting internal steel structure with perimeter wall — David Mellor Cutlery Factory, Hathersage (architect: Michael Hopkins and Partners; courtesy of Alan Brookes and Chris Grech)

10.4 Tapered beams (T-sections) penetrating glazed envelope — Hat Hill visitor's centre, Goodwood, Sussex (architect: Studio Downie)

The steel tusks of the Western Morning News building are supported by steel beams which penetrate the sloping glazed façade. A steel plate surrounds the penetration through the glazing, as illustrated in Figure 10.5. The penetration is sealed by a rubber gasket glued to the steel member.

10.5 Steelwork penetrating the façade of the Western Morning News building, Plymouth (architect: Nicholas Grimshaw & Partners)

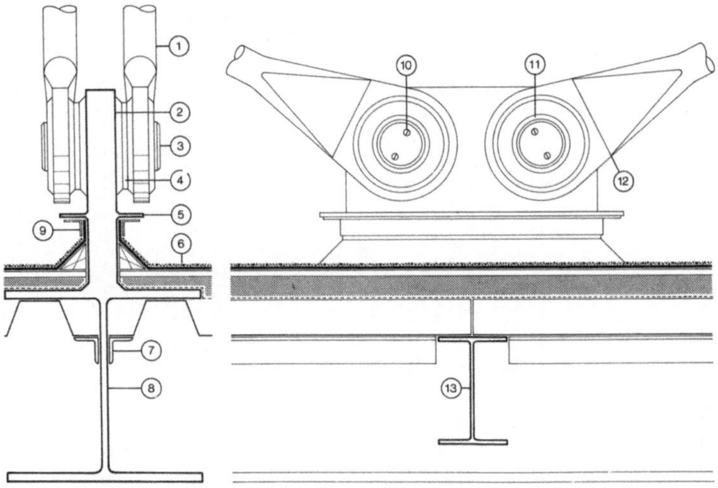

10.6 Detail of roof bracket at Sainsbury's supermarket, Camden, London (architect: Ahrends Burton & Koralek)

10.7 Architecture Enterprise Centre, Liverpool (architect: Austin-Smith: Lord)

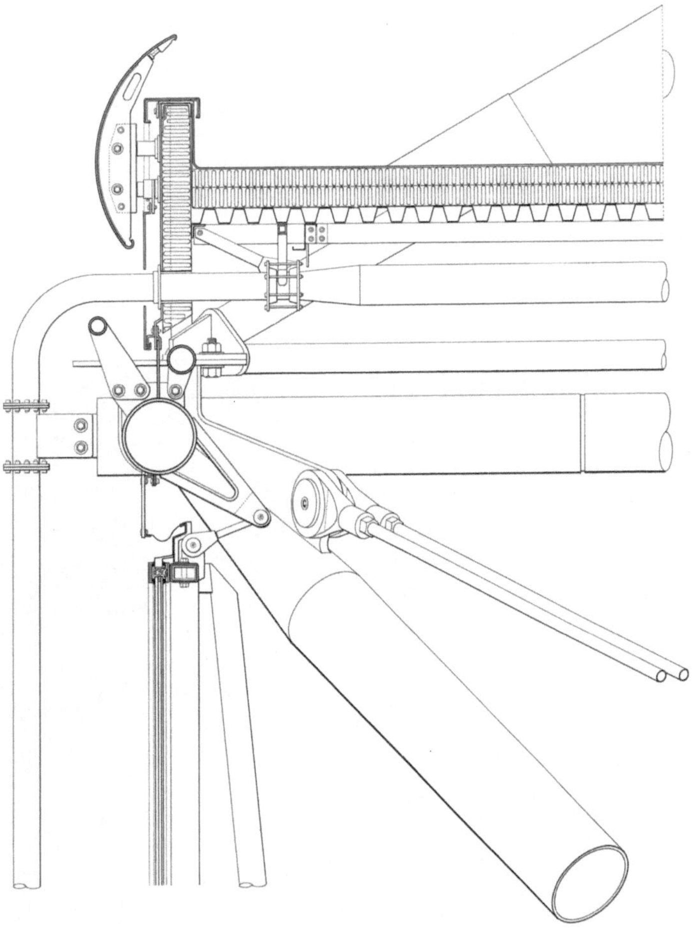

10.8 Steelwork penetrating cladding, with allowance for movement, used at Stansted Airport (architect: Foster and Partners)

10.9 Penetration through glazing at Cologne Airport (architect: Murphy Jahn Architects)

Not all penetrations are visible. Figure 10.6 illustrates the use of a bracket detail which penetrates the roof of the Sainsbury's supermarket to receive the tie-rod ends.

The Enterprise Centre at the Liverpool John Moores University is another example of where tubular members are exposed on both the inside and the outside. Arranged on a diagonal grid, with inverted tripod supports, they give the roof lateral stability without the need for heavy bracing. The tubular beams pass through the perimeter cladding and extend to support the wide overhanging eaves. The ends of the beams are propped by inclined struts as an integral part of the delicate and elegant roof assembly (see Figure 10.7).

The designer should also consider the affect of movement of the structure as it passes through the fabric. Sufficient flexibility should be included in the detail to accommodate movement of the structure and/or cladding. Figure 10.8 shows the detailing of the steelwork which penetrates the cladding and how structural movement is accommodated. Figure 10.9 illustrates this principle for a glazed façade at Cologne Airport.

Chapter 11

Technical characteristics of steel

Steel is a quality assured and dimensionally accurate product which is available in a wide range of components (see Section 1.4). The following sections present some of the key technical characteristics and design standards for steel as a material, and in its primary components. The use of stainless steel and cast steel may also be considered in expressive structures.

11.1 Specification for structural steels

Weldable structural steels are defined by British Standards BS EN 10 113,[32] and BS EN 10 025,[33] which are now the UK issues of European standards for steel as a structural material. These standards replace BS 4360, in which the most common grades were 43A and 50B, referring to the ultimate tensile strength of the steel, i.e. 430 and 500 N/mm^2 respectively. A and B refer to the toughness characteristics of the steel, which are relevant to the limit temperature at which the steel is used in service.

New designations S275 and S355JR to BS EN 10 025[33] replace grades 43A and 50B respectively, and refer to the minimum guaranteed yield strength of the material in N/mm^2. Apart from the change in designation, the properties of S275 and Grade 43 steel are the same. The yield strength of steel is approximately 70% of its ultimate tensile strength. Design resistances are based on the yield strength of steel rather than its ultimate tensile strength.

For hot-rolled tubular sections, the new standard is BS EN 10 210[34] and the most common grades are S275J2H and S355J2H, which replace the previous grades 43D and 50D respectively. J2H refers to the toughness characteristic.

For cold-rolled tubular sections, the new standard is BS EN 10 219,[35] which designates steel in the same way as for hot-rolled sections. Cold-rolled sections should not be used as a direct substitute for hot-rolled hollow sections and it is important that the standard is referred to in the designation, i.e. BS EN 10 210 - S275 J2H, or BS EN 10 219 - S275J2H.

For so-called 'weathering steels', the new standard is BS EN 10 155[36] and the common designations are S355JOWP and S355JOW, replacing grades WR50A and WR50B in the previous standard.

Technical information on the comparisons between the B and BS EN standards is available from the Corus Advisory Service (see Chapter 16 for contact details).

11.2 Design standards

The structural design of steel frames or assemblies in the UK is covered by the various parts of BS 5950. This design standard applies to the use of hot-rolled and cold-formed sections in general building construction, in which the materials are designated as in Section 11.1. BS 5950 Part 1[37] is the principal code dealing with the design of steelwork in framed structures.

UK design codes will potentially be replaced by European (CEN) standards, although BS standards are likely to continue in use for many years as they are accepted by the Building Regulations. These European Standards exist currently as 'pre-norms' or ENVs.

Eurocode 3 (ENV 1993-1-1)[38] is the relevant European design standard which covers the same application as BS 5950 Part 1.

Composite construction, which comprises steel frames and concrete or composite floor slabs, is covered by BS 5950[39] Parts 3 and 4, and, in future, by Eurocode 4 (ENV 1994-1-1).[40]

Fire-resistant design is covered by BS 5950 Part 8[41] and, in future, by ENV 1993-1-2[42] and ENV 1994-1-2.[43] However, thermal performance data of fire-protection materials is obtained from standard ISO fire tests.

11.3 Manufacturing methods for hot-rolled steel sections

11.3.1 Open sections

Hot-rolled steel I-, H-, C- and L-sections (defined in Table 1.1) are produced by continuous rolling of raw steel slabs, in a process in which the steel is sufficiently hot and malleable so that it can be formed into shapes dictated by repeated passing through multiple pairs of rolls. For I- or H-sections, internal rolls define the internal dimensions of the section and the thickness of the web, whereas external rolls control the required thickness of the flanges.

Although the rolling process is relatively simple in principle, the number of passes through the rolls and the rates of cooling are carefully controlled in order to achieve the correct dimensions and to avoid lamination problems, particularly for thicker steel flanges.

11.3.2 Manufacturing methods for hollow sections

Tubular steel is the common term given to structural hollow sections with either a square, rectangular or circular cross-section (the range of dimensions is given in Table 1.3). These sections are available in a range of sizes, weights and thicknesses. Importantly, the external dimension is constant for the range of thicknesses in a given section size. Section properties are given in Corus and SCI publications (see Bibliography in Chapter 16).

The following description of how these sections are made will help to understand how they may be used aesthetically. The two principal methods of making structural hollow sections are the seamless and the welding methods. Both methods use hot rolling to form their finished shape.

11.3.3 Seamless hollow sections

Some seamless tubes are produced by extrusion, but the majority are made by piercing the solid ingot or bar, which is then elongated in a rotary forge (called the Pilger process). In the rotary forge method, circular or tapered fluted ingots are reheated and are pierced with a mandrel using a hydraulic press. The tube is then formed by rolling externally in a helical manner, reducing the thickness and outside diameter to that of the finished tube, whilst the internal diameter remains unchanged.

The finished tube length is restricted by the weight of the ingoing billet, and the number of passes required to reduce the final size is restricted by the size of the billet. The development of continuous casting means that a greater range of billet diameter and weight can be produced, which can reduce the conversion time and also allow longer lengths to be produced.

After rolling, the tube is cropped and then passed through a series of sizing rolls which control the outside diameter. When controlled bore tubes are required, this operation is omitted since the bore of the tube is the prime dimension. Seamless hollow sections can be produced in circular form in diameters up to 500 mm and in thicknesses up to 50 mm.

11.3.4 Welded hollow sections

Welded hollow sections can be produced by a number of different methods. The most common is known as the conversion method in which round hollow sections are produced and then converted to the required shape by re-rolling. The welding can be carried out by butt or continuous welding, electric weld, spiral weld, and submerged arc welding.

A method used less commonly in the UK, but more extensively in the United States and Japan, is the direct or brake press method in which the rectangular or square section is formed by bending the parent plate and then welded. However, these sections are visually less acceptable.

Butt or continuous weld process

In the butt or continuous weld process, which is now seldom used, hot-rolled strip is heated almost to welding temperature and bent into a horseshoe shape forming a nearly closed tube. The strip edges are then heated locally and pressed together to make the weld. The hot tube then goes through sizing rolls which reduce the outside diameter to within the specified tolerance. This method is now only used for sections of up to 48 mm diameter.

Electric weld process

The majority of hollow sections used in buildings in the UK are produced by the electric weld (EW) method, which can produce circular sections from 48 mm diameter up to 508 mm diameter, square sections from 40 mm up to 400 mm, and rectangular sections from 50 × 30 mm up to 500 × 300 mm. The thicknesses produced depend on the size of the section, and range from 2.5 mm to 6.3 mm in the small sizes up to 6.3 mm to 16.0 mm in the larger sizes.

In the EW process, the strip is progressively formed into a round, nearly closed tube shape and then passes through a high-frequency induction coil which raises the strip edges to fusion temperature. The edges are then pressed together, forming a weld without the use of any filler (electrode) material. The round hollow is formed to a diameter which will create the finished section size (see Figure 11.1).

After welding, the external weld flash or bead is removed, but the internal bead is normally left untrimmed. The weld bead is regularly checked for uniformity and integrity.

11.1 Strip being formed and welded into a tube

All shapes of hollow sections can be produced as hot-finished or cold-formed sections. Both processes use hot-finished strip as their feed-stock, and are initially formed into cold round sections and then welded as described above. Cold-formed sections are then finished in their 'cold' or normal state into the desired shape. Hot-finished sections are heated and formed into a circular, square or rectangular shape whilst in the normalising temperature range. Hot-finished and cold-formed products are different in their mechanical performance and each has its own design code and product standard.

Forming the finished shape

A circular tube is the starting point for making a CHS, SHS or RHS. The circular tube is first heated and then passed through a series of rolls to produce the correct dimensions.

RHSs and SHSs are formed by passing the circular sections through a series of rolls which change the profile shape gradually into the required shape, as indicated in Figure 11.2.

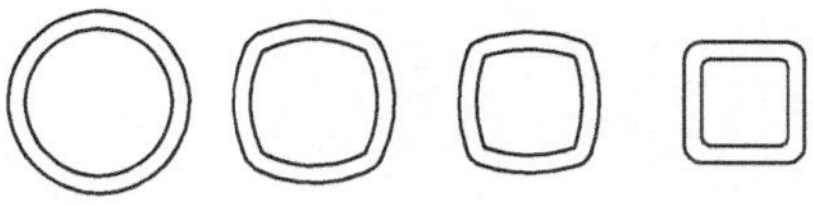

11.2 Forming the finished shape of a square section

Stretch reduction

The standard 170 mm nominal diameter tubes produced by the EW process can be stretch-reduced to produce other structural hollow section sizes up to the following nominal sizes:

- CHSs — 140 mm nominal outside diameter
- SHSs — 100 mm square
- RHSs — 120 × 80 mm.

In all cases, the maximum thickness of the section by this process is 8 mm. Stretch reduction is the main method of producing sections up to these dimensions.

Hot-finished versus cold-formed sections

Cold-formed hollow sections differ slightly in shape and form from hot-finished hollow sections. The principal differences are:

- square or rectangular cold-formed sections have larger and more rounded corner radii which may give them a less crisp appearance than hot-finished sections
- the seam weld profile is often more pronounced in cold-formed sections.

Hot-finished hollow sections are supplied to BS EN 10 210.[34] Cold-formed hollow sections are supplied to BS ENV 10 219.[35] The structural properties of cold-formed sections are less than the equivalent hot-finished sections of the same nominal dimensions. Therefore, direct substitution of cold-formed sections for hot-finished sections should not be carried out without careful checking. It is also important that the correct specification is made at the design stage.

Submerged arc welded process

This process is generally used for tubes over 500 mm diameter and up to 2134 mm diameter. Larger sizes are formed from two semicircular rolled plates, smaller sizes from a single circular rolled plate with the final weld being made by the submerged arc welding (SAW) process.

Spiral welded process
Spiral welded tubes are made by helically forming the strip which is then welded by submerged arc welding or other CO_2 processes. The method is generally used to produce large diameter thin wall tubes, for example, that are used in bored foundation piles. However, Corus no longer produces tubes by this method.

'Cigar'-shaped columns
These are made out of steel plate/cones which are welded down the seam and then in turn butt-welded onto the next piece. These are non-standard items and, as such, are relatively costly, as they have to be made to order for a particular project. A good example is shown in Figure 4.26.

11.4 Stainless steel

11.4.1 Grades of stainless steel

Stainless steel is the generic name given to corrosion and resistant steels with a minimum of 10.5% chromium, which oxidises to produce a surface patina that inhibits rusting. The stability of this thin passive layer increases as the chromium content increases, and is further enhanced by alloy additions of nickel and molybdenum.

The grades of stainless steel can be grouped into five basic groups:

- austenitic stainless steel — these are the most widely used types of stainless steels and are based on 17–18% chromium and 8–11% nickel alloys. They have high ductility and are readily weldable
- ferritic stainless steel — they contain 10.5–18% chromium, but less nickel than austenitic grades, and are less corrosion resistant
- duplex stainless steels — these steels are generally used where high stresses are to be resisted under severe corrosion conditions, and are often used for bars and pins. They typically contain 21–26% chromium, 4–8% nickel and 0.1–4.5% molybdenum
- martensitic stainless steels — these steels have a similar structure to ferritic stainless steels and are generally used for less corrosion resistant applications
- precipitation-hardened stainless steels — these steels are strengthened by heat treatment to very high temperatures and are used for heavy duty connections, such as the bolts.

Stainless steels are specified in BS EN 10088[44] and structural design is covered by supplementary rules given in Eurocode 3-1-4.[45] The most widely used grades are standard austenitic grades 1.4301 (304) and 1.4401 (316), which are specified in BS EN 10088. The designation system refers to: 1 (steel), 43 or 44 (group of stainless steels), 01, etc. (individual grade identification).

The terms in brackets refer to the equivalent ASTM (American) standard, where A304 and A316 are the common grades. The yield strength is typically 200 to 220 N/mm^2 (which is lower than for carbon steels), although the ultimate tensile strength is 500 to 700 N/mm^2 (higher than for carbon steels). Duplex steels have 400 to 460 N/mm^2 yield strength.

11.4.2 *Components*

The wide range of stainless steel components are presented in the SCI publication *Architects Guide to Stainless Steel*[31] and may be summarised as follows:

- sections in C or L or similar form that are produced by cold rolling from a flat sheet or plate
- tubular sections that are produced by bending and welding
- bars and rods
- machined pins
- brackets, often with milled serrated edges
- bolts, screws and washers.

Good examples of the use of stainless steel sections and components are:

- ledger angles for brickwork
- brackets for brickwork and cladding systems
- glazing bars and arms
- mullions to support cladding
- stair stringers and steps
- threaded bars for tie rods
- pins for tie rods
- louvres.

Various types of finish are possible, including colouring, etching, embossing and shot blasting.

A good example of the combined use of stainless steel mullions, tie rods and bracket attachments is to the glazed façade of the entrance to Tower 42, London, are illustrated in Figure 11.3.

11.5 Weathering steels

Weathering steels are a type of steel alloy in which, as weathering occurs, a hard patina is formed on the surface to prevent further corrosion. This patina can vary in colour, depending on the environment, from an orangey brown to a deep chocolatey red. A recent example of the use of weathering steel is in the cladding and structure of the Baltic Square Tower, Helsinki, shown in Figure 11.4.

Weathering steels are not suitable for industrial or coastal environments where the patina will not form properly. In the initial stages of weathering, a 'rusty' run-off may occur and adjacent areas of

11.3 Entrance lobby to Tower 42, London, showing stainless steel mullions and tie rods (architect: DEGW)

11.4 Use of weathering steel in the Baltic Square Tower in Finland (architect: Helin & Co.)

concrete or cladding should be protected. For example, where weathering steels are used for bridge girders, adequate drainage should be provided to prevent staining of concrete abutments. Where weathering steels are used for columns, gravel-filled pockets could be provided at the base to prevent staining of the concrete paving.

11.6 Use of cast steel

Steel castings have been used for many years, in particular where members meet at a 'node' point, and they offer considerable opportunities for expression. There have been significant improvements in the use of steel castings over the past two decades.

Early applications were in the Pompidou Centre, Paris, the Renault Parts Distribution Centre, Swindon, and Ponds Forge, Sheffield (Colour Plate 20).

Innovative casting technology has enabled designers to use steel castings in buildings for structural forms requiring intricate geometries, creating node connections of considerable complexity and yet possessing the essential structural quality. The avoidance of heavily-stressed welds and difficult or expensive fabrication are great advantages of castings over welded fabrication. However, the relatively long lead-in time for drawings, pattern making, casting and machining means that the use of castings should be considered at an early stage in the design process.

Castings can also incorporate fixings, such as tapped holes for threaded connections, cast-in lugs and cleats, to facilitate fabrication or erection. The development of castings in the building industry was hindered for some time by a number of misconceptions with respect to both materials and casting procedure. Castings were thought to be brittle and unweldable. This is not so, as cast steel can have similar properties to high-grade steel.

Casting design demands input from an experienced foundry, as success depends on choosing a pattern where molten metal will flow at a sufficient rate to fill the mould uniformly, and suitable non-destructive testing is needed to check that the castings do not have large inclusions or surface flaws. This is obviously most important when the castings are highly stressed.

11.5 Cast bronze supports to cladding system, Bracken House, London (architect: Michael Hopkins and Partners)

11.6.1 Examples of cast nodes and arms

Castings using steel and other alloys may be used in a wide variety of applications in steel construction, many of which have opportunities for architectural expression, as follows:

- The extended arms or 'gerberettes' of the Centre Pompidou provide the connections to the columns and tension ties. The shape of the 'gerberettes' is designed to resist combined shear and bearing forces and bending effects (see Figure 3.14).
- Ornamental details in façades, as in the bronze supports used in Bracken House, London (Figure 11.5).
- At the top or intermediate points of masts where cables connect, as used at L'Oreal, Paris (Figure 11.6).
- The nodes of tension-tie systems, particularly where a number of ties meet. In this case, the castings are designed to transfer the multiple forces smoothly (Figure 11.7).
- Connections between columns and beams, as at Bedfont Lakes near London. The use of castings is a relatively under-used method of connection in regular frames, and is generally used only in exposed applications, as in Figure 11.8 and Colour Plate 12.
- Brackets or other attachments to members.

An SCI guide[46] covers the design and use of castings. In summary, a steel casting may be preferable to other alternative forms of connection or fabrications when:

11.6 Detail of cast node at L'Oreal, Paris

11.7 Cast node between tension members at L'Oreal, Paris

11.8 Cast node between beams and columns at New Square, Bedfont Lakes, London — see also Colour Plate 12 (architect: Michael Hopkins and Partners)

- cost is not a major constraint, and where the casting or node is visually important
- relatively large numbers of components are required and the cost of welding stiffeners and other details would otherwise be excessive
- curved or softer shapes are required, e.g. saddles to column ends
- the design includes complicated tubular connections with incoming members at different angles (see following section)
- the design of the steelwork requires considerable fabrication work (steel castings can usually be made with sufficient dimensional accuracy to reduce or eliminate machining costs)
- the design uses highly tapered sections that cannot be readily fabricated
- high toughness and/or fatigue strength are important
- the connections are subject to high forces where large welds would otherwise be required.

The form and profile of the casting can reflect the structural action and forces within the connection or, alternatively, it can be designed to suit a given architectural motif or shape. The shape of the casting between the steel columns and beams in Figure 11.8 reflects the reduction of the column sizes at the floor level.

The shapes available for castings are generally only limited by the practicalities of pattern making and casting. Due to the bespoke nature of each type of casting, it is more difficult for the costs to be assessed accurately at the design stage. However, the maximum degree of economy is likely to be achieved if the casting use is repeated many times. The maximum size of an individual casting is limited only by the capacity of the foundry, and can extend to several tonnes.

11.9 Stainless steel cast legs at the Ludwig Erhard Haus, Berlin (architect: Nicholas Grimshaw & Partners)

11.6.2 *Other metals used as castings*

Iron may also be used for castings. It has a lower melting point than steel and, when molten, flows into a mould more easily. Ductile iron castings are easier to form and, as a result, are generally cheaper than the equivalent steel castings. One disadvantage is that welding to iron castings should be avoided if the iron has a higher carbon equivalent content than steel. However, bolted or pinned connections are easily made and, from the point of view of tolerance, iron is more easily machined than steel.

Castings may also be formed in stainless steel, as in Figure 11.9, which were supplied for the Ludwig Erhard Haus, the new Stock Exchange in Berlin. Bronze castings have also been used (see Figure 11.5).

11.6.3 *Castings for tubular members*

Castings are particularly suitable for connections between tubular members because:

11.10 Tubular connections of 'trunk' to 'branch' at Stuttgart Airport (architect: Von Gerkan Marg & Partners)

- the cost of profiling and welding complex tubular connections can be very high
- there are often a large number of connections of the same external dimensions in a large project
- complex hangar and node details can be formed easily
- cast connections are aesthetically more pleasing, as they are smoother in shape, and are accurate in dimensions
- toughness and fatigue resistance can be improved.

Examples of the use of castings in connections between tubular members are:

- cast steel saddles and at the head of columns
- tension connections
- noded connections in multi-planar trusses
- junctions between columns and sloping members, as in the column 'tree' of Figure 11.10
- true pinned connections, as in Figure 11.11.

11.11 Casting for pinned connections at Ponds Forge, Sheffield (architect: FaulknerBrowns)

11.12 Stainless steel node to the tubular arms at Aix en Provence TGV station (architect: AREP-SNCF and Arcora)

At Stansted Airport, the design team decided that the majority of connections could be made by castings. At Stuttgart Airport, castings were used to connect the 'trunk' to the 'branches', as illustrated in Figure 11.10.

At the TGV station in Aix en Provence, France, cast steel forked nodes connected the columns to the inclined tubular arms that supported the roof, as illustrated in Figure 11.12.

Chapter 12

Corrosion protection

The cost of corrosion protection can be as high as 20% of the total cost of fabricated steelwork, and it is therefore important not to over-specify the protective system, whilst achieving an acceptable life to first maintenance. Corus has prepared various publications[47] to assist the designer in the protection strategy. The following notes supplement the guidance given by Corus.

Corrosion of steel can only occur if both oxygen and water are present. The rate of corrosion will depend on the exposure and on the concentration of containments (usually chlorides and sulphides) in the atmosphere. Thus, permanently embedded steel piles do not corrode, even though they are in contact with water, provided that air is excluded by the impermeability of the soil. Similarly, the interior faces of tubular sections will not corrode, provided they are sealed.

Within buildings, only minor and superficial oxidation may occur, except in areas such as roofing or cladding, which may be subject to condensation or water leakage.

12.1 Internal steelwork

Inside heated buildings, where the steel is concealed from view, no protective coating is generally required. Consequently, it is now practice to use unprotected steelwork, except where the steel penetrates the envelope. For largely visual reasons, a thin coat (25 microns) of shop primer may be applied before delivery to site.

Where the steel is exposed to view internally, a decorative coat (often an alkyd) on a suitable primer (usually a zinc phosphate) may be applied. If an intumescent paint is used as fire protection, advice should be sought from the specialist paint manufacturer to ensure compatibility between it and any decorative coating.

Steelwork within perimeter walls is more susceptible to corrosion, particularly if the steel is in contact with, or embedded in, the outer leaf of a masonry wall. Bitumen, zinc-rich epoxy and pitch epoxy coatings are commonly used in such situations.

For industrial, or other more potentially severe environments, a wide range of options and coatings are available, from the use of weathering steels to metallic coatings, such as galvanizing or aluminium spray. General guidelines on the use of metallic coatings are given in BS EN 14713.[48] Provided the individual steel pieces are small, hot-dip galvanizing can be a cost-effective option.

12.2 Protective treatment specification

Guidance on the painting of steelworks is given in BS EN 12944[49] (replaces BS 5493). An adequate specification for surface protection of steelwork includes the:

- method of surface preparation
- type of protection to be used
- method of application (brush, spray)
- location of application (works or site).

The choice of specification will be determined by a number of factors:

- expected life of the structure and the maintenance plan
- environment to which the steelwork will be subjected
- whether or not the member is exposed to view
- environmental and safety considerations in both initial application and in maintenance
- size and detailing of the members
- shop treatment facilities which are available to the fabricator and/or the coatings sub-contractor
- site conditions, which will determine whether the steelwork can be coated after erection
- necessity for site-based assembly (particularly any need for welding)
- colour and surface finish or texture
- economics as influenced by the above issues.

12.3 Surface preparation

The nature of the surface to which coatings are applied has a major effect on their performance. Generally, the better the surface preparation, the better the long-term performance of the coating. Some manufacturers produce coatings which are tolerant of poorly prepared surfaces in order to facilitate maintenance and repainting. In all cases, the method of preparation should be compatible with the coating system and the manufacturers' data sheets should be consulted.

For high-quality coatings, it is essential to remove all the contaminants, rust and mill-scale, which forms when the hot surface of the rolled steel reacts with air to form an oxide. For interior environments, where a low level of protection is required, the surface should be clean and free of loose rust and mill-scale, but a high level of preparation is not generally required, depending on the system chosen.

Surfaces embedded in concrete should be free of mill-scale but may otherwise be untreated, as the alkalinity of the concrete passivates the steel. Treatment on adjacent areas should be returned for at least 25 mm within the concrete. The cover to the embedded steel should be in accordance with the requirements of BS 8110.[47]

Various methods are adopted for cleaning steelwork, but the only effective methods for removing all mill-scale and surface containments are blast-cleaning and/or pickling (immersing the steel in acid). Other methods, such as flame cleaning and wire brushing, may be useful for maintenance or preparing steel for use in mild conditions, but they will not remove all the rust and mill-scale.

The method and quality of the surface may be specified according to BS 7079,[50] some of whose parts are replaced by BS EN 8501, 8503 and 8504.[51,52] Conventional specifications may ask for shot blast to standard SA 2.5, or wire brush to standard St 2. Both grit and shot-blasting are possible, but shot-blasting is preferred for smooth high-quality paint coatings, and grit-blasting is preferred for galvanizing and some primers. Galvanized surfaces need not necessarily be grit-blasted but a thinner coating will result. Usually the manufacturer will specify the type of surface required for the application of a particular coating. Where the surface is blast-cleaned, any delay between this cleaning and the application of the first coating should not exceed four hours in order to prevent further rusting and contamination.

When site painting is adopted, it is important that the steelwork is cleaned before paint application. This may require washing of the steel with a suitable detergent to remove contamination that has occurred during transportation and erection.

For connections exposed to severe environments, it may be necessary to blast-clean the connections before applying the protective system, but this is time consuming and expensive. Alternatives include the use of galvanized bolts, de-greased after tightening, followed by etch priming and painting to the same specification as the adjacent surfaces.

12.4 Type of protection to be used

12.4.1 General

The following notes may assist in determining the type of system which is appropriate for a particular application and exposure:

- Protection requirements are generally minimal inside dry, heated buildings, such as offices, shops and schools. In such situations, no protection is required, except for decorative reasons, unless the steelwork is located within a cavity wall (see below).
- Some interior steelwork may be exposed to condensation, particularly in sports halls, exhibition halls and workshops. Although these environments are generally classified as mild, a protection system will generally be required for visual reasons.

- Where steelwork is exposed to industrial processes, or encloses humid areas such as swimming pools, the environment should be classed as 'severe' and specialist advice should be sought. The system chosen will depend on the particular environment and atmospheric contaminants.
- Perimeter steelwork enclosed in cavity walls can be subdivided into two categories:
  - where an adequate air gap (40 mm minimum) exists between the steel and the outer masonry skin, then adequate protection can be achieved by a simple painting system. This gap can be reduced to 25 mm if an impermeable thermal insulation is placed between the steelwork and the outer leaf. Suitable materials include closed cell organic foams, such as polyurethane phenolic based or extruded polystyrene
  - where the steelwork is in contact with, or embedded in, the outer leaf of the wall, then a high-quality surface protection system should be applied. Bitumen-based coatings (site applied) and coal tar epoxies (shop applied) are common. Alternatively, for severe environments, the steel can be galvanized and painted with bitumen-based paint to prevent the galvanized surface coming into contact with the masonry.
- Where the steelwork is fire protected, consideration must be given to the compatibility between the corrosion and fire-protection systems. For boarded fire-protection systems, this will generally not be a problem. Mineral fibre or particle-sprayed systems should generally not be used on galvanized steelwork due to adhesion problems. Where intumescent paints are used, it is vital that the compatibility is checked with the paint manufacturer.

12.4.2 Paint coatings

There is a large choice of painted coatings suitable for different environments and methods of application. Specific guidance on any particular type of coating is outside the scope of this publication, but detailed guidance can be obtained from manufacturer's catalogues and data sheets. The guidance given by Corus[47] concentrates on a few generic types of system suitable for various applications and environments. This information should be consulted as a source of information against which manufacturers' data can be compared.

In order to make rational decisions on the type of coating system to be used, it is necessary to understand a few basic concepts about paints and paint systems, as follows. Paints consist of three basic components:

1. The *pigment* — pigments are fine organic or inorganic compounds which provide colour, opacity, film cohesion and sometimes corrosion protection. For example, zinc and micaceous iron oxide (MIO) coatings provide significant levels of corrosion protection.

2. The *binder* — binders are usually resins or oils but can be inorganic compounds, such as soluble silicates. The binder is the film-forming component in the paint.
3. The *solvent* — solvents are used to dissolve the binder and to facilitate application of the paint. They are usually organic liquids or water. As the paint 'dries', the solvent evaporates into the atmosphere.

Until recently, rapid drying paints contained solvents, such as volatile organic compounds (VOCs). Recent environmental legislation has required paint applicators to control the amount of VOCs which are released. Paint manufacturers are therefore beginning to produce 'high solids' paints (i.e. low solvent) paint and water-based paints which do not release harmful compounds.

As the solvent evaporates, the thickness of the film reduces. The thickness of the paint coating may therefore be referred to by both its 'wet' and 'dry' film thickness. The wet film thickness can be measured by a comb-gauge allowing immediate correction of the film thickness. The dry film thickness can be measured by the use of various types of magnetic or electro-magnetic gauge. Paint thicknesses given in specifications are usually the final required dry film thicknesses.

Paint classification — the most common method of classification of paints is by their pigmentation or their binder type. Zinc phosphate primer, for example, may be used with an alkyd, epoxy or acrylated rubber binder. Care should therefore be exercised when specifying the type of paint, as the performance of zinc phosphate epoxy will be significantly different to that of a zinc phosphate alkyd or zinc phosphate acrylated rubber.

Paint systems — a full painting system will often include a primer, an intermediate coat and a finishing coat. The cost of application of the paint can be a significant factor and the tendency is to use as few coats as possible. In some cases, the same paint can be used for all three coats and the intermediate coat and finishing coat applied as one high build (HB) coat. In all cases, the various coats of paint must be compatible with each other, and the manufacturer should be consulted in order not to invalidate warranties.

The following additional points should be considered in the choice of paint system:

- The paint system should be compatible with the available method of surface preparation and the method of application. For example, site-applied systems should not generally require that the surface is grit-blasted, or that the coating is applied by spray.
- The environmental impact of the coating system should be considered. Paint systems which were once very popular, such as chlorinated rubbers, are being phased out due to environmental considerations during manufacture. Paints with low levels of VOCs are increasingly popular.
- Where paint systems are site-applied, the safety of the operatives must be considered. Expensive mobile working platforms (MWPs) or scaffolding may have to be used. Generally, brush

application is preferred on site, as sprayed coatings usually have to be applied in a contained environment and the operative may require air-fed respiratory equipment.

- Two pack epoxies have poor resistance to ultraviolet radiation and are highly susceptible to 'chalking'. Over-coating problems can occur when epoxies are applied in two coats, unless they are over-coated before the first coat is fully cured. This is particularly important where a two coat system is partly applied in the shop and partly on site. A 'travel coat' may be required as an intermediate coat between the two epoxy layers. Re-coatable epoxies are becoming more common.
- Prefabrication primers generally have to be applied in the workshop within four hours of blast cleaning. They are usually applied in thin films, in the order of 25 microns (10^{-6} m) thick, and their durability is limited. Many modern primers based on synthetic resins are not compatible with manually prepared surfaces as they have a low tolerance to rust and mill-scale. Conversely, many oil- and alkyd-based primers, which are tolerant to hand-prepared surfaces, cannot be over-coated with finishing coats that contain strong solvents, such as acrylated rubbers, epoxies and bituminous coatings.

12.4.3 Hot-dip galvanizing

In the galvanising process, the steel is first cleaned of rust and mill-scale by blast cleaning and pickling (by dipping in dilute hydrochloric acid containing a rust inhibitor). It is then dipped into a bath of molten zinc at a temperature of about 450°C. At this temperature, the steel reacts with the molten zinc to form a series of zinc/iron alloys on its surface. As the steel element is removed from the bath, a layer of relatively pure zinc is deposited on top of the alloy layers and then solidifies as it cools, often assuming a crystalline metallic lustre, usually referred to as 'spangling'. This has been exploited by some architects as an attractive finish in its own right. One example, which won the annual architectural Galvanizers' Association award, is a staircase at the Fruit market in Edinburgh, shown in Figure 12.1.

The thickness of the galvanized coating in the hot-dip process is influenced by various factors:

- The size of the element — thicker, heavier sections tend to lead to thicker coatings.
- The surface roughness — blast-cleaned surfaces tend to lead to thicker coatings.
- The steel composition — the amount of silicon used in high-strength steel can have a marked affect on the coating weight deposited.

'Hot-dipping' is a bath process and there is obviously a limitation on the size of components which can be galvanized. Double-dipping can often be used when the length of the workpiece exceeds the size of the bath. Specialist companies should be consulted if the length of

12.1 Staircase at the Fruit Market Gallery, Edinburgh (architect: Richard Murphy)

the member exceeds 5 m. The length limitation often makes galvanizing unsuitable for the treatment of primary beams and columns in multi-storey buildings. Some distortion of lighter fabricated members can be caused by differential thermal expansion and contraction, and by relief of residual stresses in the member.

The specification of hot-dip galvanized coatings for structural steelwork is covered by BS EN ISO 1461[53] (replaces BS 729). For sections greater than 5 mm thick, a minimum average zinc coating weight of 610 g/m^2 is required, equivalent to a zinc coating of 85 microns thickness. This coating will offer sufficient protection for over 30 years in a clean rural environment, and between 10 to 25 years in marine and urban situations.

Newly galvanized surfaces are difficult to paint because of adhesion problems, unless an etch primer or mordant wash is used. Weathered galvanized surfaces are more tolerant, provided that a suitable paint system is used. Connections between galvanized components should generally be bolted rather than welded, due to the poor quality of welded joints and the fumes produced during the welding process.

12.4.4 Galvanizing of tubular members

Certain special requirements for galvanizing tubular sections are noted below:

Table 12.1 Suitable sizes of vent holes in tubular member

Size of tubular section (mm)	Minimum diameter of hole (mm)
< 25	8–10
25–50	12
50–100	16
100–150	20
> 150	25

- Filling, venting and draining should be provided in tubular members. Holes not less than 10 mm diameter must be provided in sealed tubular sections to allow access for molten zinc, venting of hot gases and the subsequent draining of zinc.
- Sealed tubes which are to be galvanized will require vent holes to allow the escape of hot air during the hot-dipping process. Table 12.1 indicates suitable hole sizes.

Generally, tubular sections should not be welded after galvanizing due to the possibility of poor welding due to porosity caused by the gases given off from the zinc coating. Sections are generally welded before galvanizing and components bolted together after galvanizing.

Owing to the smooth surface created by galvanizing, a special etching primer or 'T' wash or light 'sweep' blasting will be required before painting, before the final two coats are applied on site.

12.4.5 Metal spraying

An alternative method of applying a metallic coating to steelwork is by spraying. In this case, either zinc or aluminium can be used. The metal, in powder or wire form, is fed through a special spray-gun containing a heat source which can be either an oxy-gas flame or an electric-arc. Molten globules of the metal are blown by a compressed air jet onto the previously blast-cleaned surface. No alloying occurs and the coating, which consists of overlapping platelets of metal, is porous. These pores are subsequently sealed by applying a flood coat of clear or pigmented epoxy or polyurethane coating. Further painting is then optional.

The adhesion of sprayed metal coating to steel surfaces is considered to be essentially mechanical in nature. It is therefore necessary to apply the coating to a clean roughened surface; blast-cleaning with a course grit abrasive is normally specified. This would usually be chilled-iron grit, but for steels with a hardness exceeding 360HV, alumina or silicon carbide grits may be necessary.

Coating thickness varies between 100 to 250 microns for aluminium, and 75 to 400 microns for zinc. Both metals perform similarly in most situations, but aluminium is more durable in highly industrial environments.

Metal spray coatings are usually applied in the fabrication shop. Unlike hot-dip galvanizing, there is no limitation on the size of the element and, as the steel surface remains cool, there are no distortion problems. Relative costs will vary depending on the size of the section, but metal spraying might typically be twice as expensive as galvanizing.

The protection of structural steelwork against atmospheric corrosion by metal-sprayed aluminium or zinc coatings is covered in BS EN 22063.[54]

12.4.6 *Electroplating and sheradising*

Electroplating and sheradising are processes which are used for the application of a metallic coating to small components, such as fittings and fasteners. Due to the thin coating that is applied, they should generally only be used for mild environments, or for decorative purposes.

12.4.7 *Common protection systems*

The protection systems detailed in this section (see Tables 12.2 and 12.3) are based on those which are currently recommended by Corus for various environments.[47] The system numbers are those allocated by Corus and the recommended paint thicknesses are the dry film thickness. These thicknesses may vary between manufacturers.

Systems have been chosen which are suitable for the following environments:

- *Interior*:
  - *low risk*: where there is a low risk of condensation, exhibition halls, workshops, sports halls
  - *medium risk*: production buildings, or those with potentially high humidity and some risk of pollution.
- *Exterior*:
  - *medium risk*: most rural and urban areas, with low sulphur dioxide, acid, alkali and salt pollution
  - *high risk*: urban and industrial atmospheres with moderate sulphur dioxide and/or coastal areas with low salinity
  - *very high risk*: industrial areas with high humidity and aggressive atmospheres. Coastal and offshore areas with high salinity.

The tables show both interior hidden and interior visible, to facilitate the choice of one system in structures where parts of the steelwork are hidden and parts are visible.

The tables do not cover the following:

- *Interior*:
  - *very low risk*: with a dry atmosphere (e.g. offices, shops, schools, etc.) as, generally, no protection system is required, except for decorative reasons.

Table 12.2 Recommended corrosion protection systems for interior environments[47] ($\mu m = 10^{-6}$ m)

Environment	Interior low risk		Interior medium risk						
	Hidden	Visible	Hidden	Visible	Hidden	Visible	Hidden	Visible	Visible
Structure life in years[(1)]	55+	55+	Unknown	Unknown	60+	60+	55+	45+	40+
Coating life in years[(2)]	N/A	15	N/A	Unknown	N/A	20+	N/A	25	20
BS system number	B3	B10	B4	B11	B6	B13	B7	B14	B15
Shop applied surface preparation	Blast clean to SA 2.5	Blast clean to SA 2.5	Blast clean to SA 2.5	Blast clean to SA 2.5			Blast clean to SA 2.5	Blast clean to SA 2.5	Blast clean to SA 2.5
Coating	Zinc phosphate epoxy primer (80 μm)[(3)]	Zinc phosphate epoxy primer (80 μm)[(3)]	Water-based acrylic or epoxy zinc phosphate primer (2 × 60 μm)[(4)]	Water-based acrylic or epoxy zinc phosphate primer (2 × 60 μm)[(4)]	Hot-dip galvanize to BS 729 (85 μm)[(5)]	Hot-dip galvanize to BS 729 (85 μm)[(5)]	High solid epoxy zinc phosphate primer (80 μm) High build recoatable epoxy MIO (120 μm)	High solid epoxy zinc phosphate primer (80 μm) High build recoatable epoxy MIO (120 μm)	High solid epoxy zinc phosphate primer (100 μm) High build aliphatic polyurethane finish (120 μm)
Surface preparation	None	Wash free of contamination	None	Wash free of contamination	None	Mordant wash	None	Wash free of contamination	None

Notes:

1. Structure life — the number of years of freedom of severe corrosion that might lead to weakening of the structure.
2. The expected number of years to first maintenance of the coating.
3. Can be applied as a 20 μm prefabrication primer plus 60 μm post-fabrication primer.
4. Water-based technology is still developing and advice should be obtained from the manufacturer.
5. A thickness of 85 μm can be achieved on steel over 6 mm thick without grit-blasting.

Table 12.3 Recommended corrosion protection systems for exterior environments[47] ($\mu m = 10^{-6}$ m)

Environment	Exterior — medium risk			Exterior — medium risk		Exterior — very high risk	
Coating life in years[(1)]	20+	20	20	25	25	20	25
BS system number	B12	B14	B15	E6	E9	E9	E10
Shop applied surface preparation		Blast clean to SA 2.5	Blast clean to SA 2.5	Blast clean to SA 2.5	Blast clean to SA 2.5	Blast clean to SA 2.5	Blast clean to SA 2.5
	Hot-dip galvanize to BS 729 (85 μm) High-build epoxy MIO (100 μm)	Zinc phosphate epoxy primer (80 μm) High-build epoxy MIO (100 μm)	High solid epoxy phosphate primer (100 μm) High solid aliphatic polyurethane finish (100 μm)	Zinc-rich epoxy primer (100 μm) High-build epoxy MIO (100 μm)	Zinc-rich epoxy primer (100 μm) High-build epoxy MIO (200 μm)	Zinc-rich epoxy primer (100 μm) High-build epoxy MIO (200 μm)	Zinc-rich epoxy primer (100 μm) High-build epoxy MIO (200 μm)
Surface preparation	None	Wash free of contamination	None	Wash free of contamination	Wash free of contamination	Wash free of contamination	Wash free of contamination
Coating	None	Recoatable polyurethane finish (60 μm)	None	High-build epoxy MIO (100 μm)[(2)]	High solid aliphatic polyurethane finish (60 μm)	High solid aliphatic polyurethane finish (60 μm)	Recoatable polyurethane finish (60 μm)

Notes:

1. Coating system durability is based on experience and is the expected life in years before major maintenance. It is not a guaranteed life expectancy.
2. It should be noted that the available colours are limited.

- *Interior*:
  - *high risk*: buildings with high humidity and corrosive atmospheres, e.g. chemical plants, swimming pools, paper manufacturing plants, etc. In these cases, it is recommended that specialist advice is sought.

For steelwork in contact with a masonry outer skin it is recommended that the steel is galvanized or, alternatively, it is blast-cleaned and one coat of solvent-free epoxy (dry film thickness of 450 μm) is applied. If the steelwork is galvanized, two coats of heavy duty bitumen (total dry film thickness of 200 μm) should be applied.

12.5 Method and location of application

Spraying is the most widely adopted method for application of paint systems in a workshop with a controlled environment, often with forced ventilation. Both air-fed and airless spray are possible. Airless spraying is now more common, as application rates are higher and overspray is reduced. Brush application is usually carried out on site, where control of the local environment is more difficult.

Site coating of previously primed steelwork is often preferred by steelwork contractors as it allows the steelwork to be moved rapidly out of the workshop and on to site. The other main advantage is that, as the final coat will be applied on site, repair of damage caused by transport and erection is not required and the finished appearance will be more uniform.

However, a number of important aspects should be considered, as noted earlier:

- the quality of the site-applied finish coat will be inferior to that applied in the workshop, particularly if the atmosphere is polluted or in coastal regions (both chlorides and wind-blown sand, for example, can penetrate a considerable distance inland)
- all paintwork should be cleaned on site, before application of the site coat
- on-site coatings should generally be applied by brush
- damage to paint systems should be minimised by the careful use of strops, wrapped chains or lifting lugs.

The steelwork for the Igus Factory in Germany was blast-cleaned and prepared with its primary coating before being transported to site. The final protective coating was applied on site (see Figure 12.2).

12.6 Protection of connections

The method of protecting the connections depends on the type of connection and the system used for the steelwork. In severe environments, galvanized bolts may be used.

12.2 The Igus Factory, Cologne (architect: Nicholas Grimshaw & Partners)

Normal bearing-type bolted connections require protection of the contact surfaces. Although a priming coat is sometimes recommended, it is worthwhile to treat the contact surface to at least the same standard as the main steelwork, as it is unlikely that these surfaces will be repainted. In extreme cases and over a long period, corrosion between the surfaces can lead to lamination of the steel and failure of the connection.

For high-strength friction grip-bolted connections, the faying or contact surface should be free from any containment or coating which would reduce the slip factor required in the connection. Most paint systems will be detrimental to the frictional resistance, but some inorganic zinc silicate primers and spray metal coatings can increase the slip factor, and, hence, improve the resistance of the connection to shear force. Where required, the faying surfaces should be masked to prevent painting of the surfaces. All high-strength friction grip (HSFG) connections should be sealed round their edges when the bolts have been tightened. After making the connection, the area should be painted to repair any damage caused to the paint system, and to protect the exposed part of the connection.

12.7 Detailing of exposed steelwork to reduce corrosion

General guidance on factors relevant to the corrosion protection of exposed steelwork are as follows:

- Avoid sharp edges, sharp corners, cavities, crevices, etc. Edges and corners can be smoothed to improve the adhesion of the coating. It is usual to apply an additional 'stripe coat' to edges and welds.
- For corrosion protection, welded connections are preferred to bolted connections. Butt-welded connections are preferred to lap joints, and continuous welds are preferred to intermittent welds which can create pockets in which rusting could occur.
- Lap joints should be orientated so as to prevent build-up of water or contaminants.
- Drainage holes should be provided where necessary, but do not allow water to run on to other parts of the structure.
- Seal 'box' sections. If box girders are fully sealed, corrosion of the interior steel surface will not occur. Where it is difficult or impracticable to obtain a complete seal, desiccants such as silica-gel can be used, at the quantity of about 250 g/m^3 of void. This process is effective for two to three years. Tubular sections, which are sealed during fabrication, need no internal treatment.
- Provide free circulation of air around the structure. This increases the rate of drying of the surfaces. Premature breakdown of protective coatings can occur in sheltered areas where moisture remains in contact with the coating for long periods. In such situations, additional protective layers should be considered.

- Eliminate dead-flat surfaces and crevices where debris and water might accumulate. In addition, suitable accessibility for routine maintenance and inspection should be provided in highly sensitive locations.
- At foundation level, steel is more susceptible to corrosion and damage. For example, a concrete plinth with a sloping top can be provided, although this may not be architecturally appropriate in all circumstances.
- Waterproofing of exposed concrete slabs is important to keep water from penetrating to the steel beneath. This is important in car parks.

12.8 Contact with other materials

General advice on contact with other materials is given as follows:

- Avoid connections between different metals or, alternatively, insulate the contact surfaces. Materials such as zinc and aluminium offer sacrificial protection, whereas others such as stainless steel, could cause accelerated corrosion of carbon steel at the contact surface.
- Provide an adequate depth of cover and quality of concrete to encased sections.
- Separate steel and timber by the use of coatings or sheet plastics (many woods are potentially corrosive due to their treatments).

Chapter 13

Fire protection

Fire resistance is a requirement of all national Building Regulations in order to ensure stability of the structure in the event of fire, to prevent fire spread, to allow means of escape of the occupants and also for the safety of fire fighters. Although not stated explicitly in regulations, fire-resistance measures also ensure protection against disproportionate damage, so that small fires will not require major structural repair. Fire resistance requirements are a function of:

- the size of the building (especially the height)
- the fire load (combustible contents)
- the presence of 'active' fire protection measures, e.g. sprinklers.

A holistic 'fire engineering' approach can be used to predict the severity of fires, and their effect on the primary structure. Fire engineering is often used to justify the use of unprotected steel structures in airport terminals and sports halls, etc., where fire loads are low and the means of escape are good.

The following sections present the strategies that may be employed to ensure the fire safety of steel structures, depending on whether the steelwork is concealed (i.e. is not visually important) or is exposed to view.

13.1 Forms of fire protection

The SCI and the Association of Specialist Fire Protection Contractors and Manufacturers Ltd (ASFP) have produced detailed guidance[55] listing the majority of approved materials and presenting tables of protection thicknesses for various periods of fire resistance.

In principle, the main forms of fire protection to steel members that may be considered, are as follows:

- Sprayed protection around the profile of the member. Sprayed materials may be of a variety of forms, including, for example, cementitious materials and vermiculite-cement.

- Board protection as a box around the member. Board protection can be fixed on secondary noggins or, in some cases, to itself to form the box. Board protection is preferred for columns because it forms the finished shape of the column. Softer forms of protection, based on mineral fibre, are also available for beams and trusses.
- Wrapping in a fire-protective layer and sheathing by pressed metal panels to preserve the architectural profile of the member.
- Concrete encasement of I-sections.[56] Partial encasement between the flanges of I-sections can achieve good fire resistance[24] and can be introduced as an off-site process.
- Intumescent coatings, which expand on heating to provide an insulating layer. Essentially these coatings have the appearance of paint, but thicker coatings can have a fibrous appearance.
- Concrete-filling of tubular columns, including the provision of additional reinforcement for larger periods of fire resistance.
- Water-filling of tubular sections, as part of a continuous tubular structure, such as a roof truss or an external structure.

Essentially, the choice of fire-protection method depends upon a number of factors, summarised as follows:

- Initial cost of application.
- Appearance, if exposed to view.
- Durability and life to first maintenance.
- Compatibility with the environment (during construction and in service).
- Compatibility with the corrosion-protection system.
- Size and shape of the member.
- Implications for the construction programme.
- Avoidance of mess and dust during installation.
- Ability for the protection to be applied off-site.
- Maintenance and repair after damage.

Fire-protection systems are generally assessed on a failure temperature of 550°C, which corresponds approximately to 60% of the original strength of the steel member. Some manufacturers present design tables for a range of failure temperatures, which may be useful when selecting the fire protection required for lightly stressed members.

13.2 Sprayed and board protection

Sprayed protection is applied manually by spraying around the profile of the section, and is a potentially messy operation, which results in a rough surface texture. Therefore, sprays are generally used in concealed applications, such as beams with suspended ceilings. A typical sprayed protection to cellular beam is shown in Figure 13.1. They are cost-effective and a relatively fast site operation. For most applications, a spray thickness of 20 to 30 mm would be suitable.

Board protection is of two types: hard boards and softer boards. Hard boards, such as fire-resistant plasterboard, can be used as a

13.1 Sprayed protection to cellular beams

facia, whereas softer boards are usually concealed. Boards provide a 'box' protection around the member, and are held in place by a light framework. The enclosed box is usually 70 to 100 mm wider than the steel section.

13.3 Intumescent coatings

Intumescent coatings are applied as paints and are preferred where the steelwork is exposed to view (Figures 9.2 and 13.2). They expand on heating to form an insulating layer which protects the steel section. There are two basic types of these coatings:

- thin-film coatings (<2 mm thick) — these coatings can easily achieve 60 minutes' fire resistance and, for beams and columns, 90 minutes. They are generally appropriate for internal applications

13.2 Use of intumescent coating to a curved portal frame structure at Cheltenham Racecourse

- thick-film coatings (3 to 6 mm) — these coatings can provide up to 120 minutes' fire resistance (or longer in some cases). They are often appropriate for external applications.

Thin-film coatings are a particular type of coating and not a thinner film of a thicker system. Thin-film coatings retain the basic shape of the steel section, whereas thick-film coatings can appear relatively coarse, and may have an 'orange peel' texture. Finishing coats may or may not be applied on site. Some fabricators prefer on-site application rather than factory-applied coatings, as handling on site may cause some damage. Furthermore, touching up of the surface does not achieve the same quality of finish as a factory applied coating.

In the factory, the paint would be sprayed, whereas on site it would be applied by brush or roller, which has a different visual effect. Architects generally take the view that on-site painting is preferred to ensure a consistent finish. However, on-site application is more expensive, and the control of thickness and the prevention of contamination can be difficult.

Off-site application of intumescent coatings has been used on a number of recent projects and is gaining popularity because of its implications on a reduced construction programme, particularly in multi-storey buildings (Figure 4.11). Guidance on the use of off-site intumescent coatings is given in an SCI publication.[57] They are applied as a single coating of up to 1.5 mm thickness in order to speed up the drying operation. A high level of quality control can be achieved in off-site processes (see Figure 13.3).

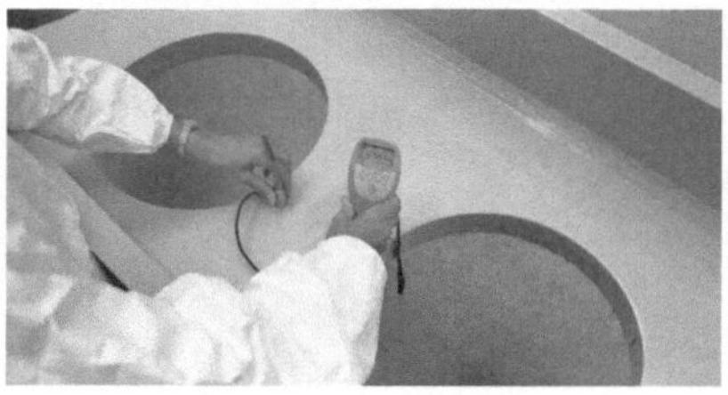

13.3 Intumescent coating to cellular beam being measured for its coating thickness

The exposed area of tubular sections is considerably less than for open sections, such as I-beams. Therefore, tubular sections heat up less quickly in a fire, and so require less fire protection to achieve a specific fire resistance. Guidance is given in the ASFP/SCI publication.[55] SHS or RHS require a slightly greater coating thickness than CHS because of possible cracking of the intumescent coating at the corners of the section.

13.4 Partial encasement by concrete

Many forms of steel construction have been developed, such as *Slimdek*[15] and partially encased columns,[24] which do not require additional fire protection.

In *Slimdek*, only the bottom of the beam flange is exposed, and the remaining part of the steel section is encased by the concrete or the slab. In fire conditions, the upper part of the section resists the applied moment, despite the much reduced strength of the bottom flange. The range of the Asymmetric *Slimflor* Beams (ASB) section has been designed to achieve 60 minutes' fire resistance. Additional protection may be applied to the bottom flange for longer fire-resistance periods.

Partially encased columns are formed by placing concrete between the flanges (usually as an off-site operation). Bar and link reinforcement may be attached to the steel section. In fire conditions, the flanges of the column (or beam) are exposed but the steel web, concrete and bar reinforcement remain relatively cool and resist the applied compression. A fire resistance of 60 or 90 minutes can be achieved, depending on the amount of reinforcement provided.

Fully encased columns are rarely used in modern steel construction, except where durability or resistance to impact is a concern, or a long period of fire resistance is required. The size of the encased column is usually 150 to 200 mm wider than that of the steel section.

13.5 Concrete filling of tubular sections

Tubular steel columns can be filled with concrete,[25,43] to provide up to 120 minutes' fire resistance with additional bar reinforcement. Concrete filling is carried out on site and so the fast erection times of steel construction are maintained.

The advantages of using concrete-filled tubular sections are:

- the steel section dispenses with the need for formwork and supports the loads during construction
- the erection schedule is not dependent on the concreting operation or curing time
- the columns are slender and possess good compression resistance
- additional external fire-protection is not necessary for up to 60 minutes' fire resistance
- if required, fire protection can be added later to increase the fire resistance.

Concrete filling of tubular sections requires no special equipment and the filling operation may be integrated into other concreting operations. However, it is necessary to provide ventilation holes in the column walls to prevent the dangerous build up of steam pressure inside the column in the event of a fire. Two full diameter holes placed diametrically opposite each other, both at the top and bottom of each storey height, have been used in testing and have proved to be adequate. The holes are positioned outside the level of any floor slab or screed. A drain hole should also be provided at the base of a column to prevent water collecting if it is left standing empty on site prior to filling.

13.4 Cheung Kong Tower, Hong Kong with large diameter concrete-filled tubular columns (architect: Cesar Pelli & Associates Inc.)

For large-diameter tubular columns, concrete may be pumped from the base of the column through a valve, as was done at Hong Kong's Cheung Kong tower (Figure 13.4). The valve is later cut away when the concrete has gained adequate strength. Two or three storeys may be filled in this method. A detail is illustrated in Figure 13.5.

Considerable research has been undertaken into the structural and fire-resistance performance of concrete-filled hollow sections, aimed at developing design procedures for this form of construction. The results of recent research in the subject is incorporated into national and international codes of practice. Calculation methods are included in Eurocode 4 Part 1.2.[43] The fire resistance aspects of concrete-filled sections are also covered in BS 5950 Part 8.[41]

13.5 Valve used for concrete filling of large diameter tubular columns

13.6 Water filling of tubular sections

Water filling is often mentioned as a method of providing fire resistance to tubular structures, but is rarely carried out in practice. The earliest and most well-known example is the external structure to Bush Lane House in Cannon Street, London (see Figure 13.6). At the new Cargo Handling Facility of Hong Kong Airport, the long-span tubular trusses were water filled to provide 120 minutes' fire resistance, and the prototype trusses were fire tested to prove their effective performance. These curved trusses are illustrated in Figure 13.7. The pre-requisites of this system are a continuous passage for water through the tubular structure, which is itself problematical in structural design terms, and the need for a water replenishment system, which is fed by gravity. Guidance is given in an SCI publication.[58]

13.6 Water-filled tubular façade members at Bush Lane House, London (architect: Arup Associates)

13.7 Curved water-filled roof trusses at the HACTL Superterminal One Cargo Handling Facility, Hong Kong International Airport (architect: Foster and Partners)

13.7 Fire protection by enclosure

If the steelwork is exposed, protection by an additional enclosure is probably the most common form of protection and, with open sections, it adopts the rectangular or square form of the member it encloses. Encasing tubular steel, particularly CHS sections is more problematical, yet architecturally the results do not necessarily detract from the original form.

At Kansai International Airport, Renzo Piano clad the cigar-shaped columns with a glass reinforced cement (GRC) fireproofing which provided 60 minutes' fire resistance. The plasticity of the GRC was used to give the columns soft and organic forms, and to complement the exposed steelwork. Examples of these details are shown in Figure 13.8.

13.8 Fire engineering

As the nature of fire protection of buildings and the safe evacuation of occupants becomes more complex, a fire-engineering approach can be undertaken which may justify the use of unprotected steelwork, provided other 'compensatory measures' are incorporated, such as sprinklers/mechanical smoke extract systems, etc. This is

13.8 Fire clad columns, Kansai Airport (architect: Renzo Piano)

particularly important for buildings in which traditional fire protection is aesthetically or functionally not desirable, such as in sports stadia, railway stations, large halls, etc.

The standard fire-resistance test, and the tables of requirements in the Building Regulations which relate to it, provide the means of achieving satisfactory performance of structures in fire. Conditions during a real fire, however, differ from those during a standard fire test. Methods have been developed which enable the behaviour of structures in real or 'natural' fires to be predicted with greater accuracy.

It is possible to demonstrate that internal steelwork may be designed as unprotected in cases where:

- the buildings have low fire load
- the steelwork supports only the roof
- active fire protection measures are installed — these measures may include fire detection devices, sprinklers, and/or other smoke control systems.

Recent full-scale fire tests on an eight-story composite steel framed building at BRE Cardington have demonstrated that composite beams may be designed as unprotected in many cases for buildings with up to 60 minutes' fire resistance.[59] This good behaviour in fire mainly occurs due to tensile membrane action in the floor slab, which is a property of composite structures.

The 'fire engineering' approach would normally be carried out by a fire engineering consultant, who, in broad terms, would consider the overall risks and effects of a fire on the structural stability. An overall strategy would be developed to deal with this fire scenario. It differs from the normal design methods, requiring fire protection for the individual elements of structure, based on a notional failure temperature.

The fire engineering method can be divided into four main stages:

1. Determination of the fire load (expressed as kJ/m^2 floor area or kg/m^2 of wood equivalent).
2. Prediction of the maximum fire temperature.
3. Prediction of the maximum steel temperature.
4. Assessment of the structural stability (including loss of bracing members).

Fire engineering is not generally used for small buildings where the expense of more detailed analysis is unjustified. Nor is it an approach that can be used for buildings which are subject to a change of use, such as advanced factory units, especially where occupancies are not 'fixed'. It can be advantageous in structures such as sports halls or airport terminals where the change of use is unlikely and the fire load is low and predictable.

13.9 External steelwork

Steelwork that is external to the building envelope can be designed as unprotected,[60] if the heat emanating from the windows or other openings in the façade does not cause the strength of the steel to reduce to a point where it cannot support the applied loads. Architecturally, this is less of a problem for external columns than for external beams, which are often in the direct path of the flame energy from a window below.

Various design methods exist for external steelwork subject to direct flames or radiant heat. These methods have been used to justify a wide range of structural designs. Additional measures, such as shielding of the beam, can also be introduced. These approaches are normally more appropriate for buildings with discrete windows in heavy weight façades, such as brickwork, rather than continuous windows in lightweight façades.

The temperature of the exposed steel would normally be designed to be below 600 °C in order that it possesses reasonable strength retention for structural efficiency under the reduced loads in fire conditions.

Chapter 14

Site installation

The following sections refer to practical aspects of steel construction, including bolting, welding, deflections and allowance for tolerances.

14.1 Bolting

A limited range of bolt diameters and lengths is usually detailed on a given project, depending on the plate or element thickness to be joined. Fully threaded bolts are gaining popularity as they can be used with a range of plate or member thicknesses.

Countersunk holes and bolts are often appropriate for use on the outside faces of columns or beams to avoid protrusion of the bolt heads, although fabrication using countersunk bolts is more expensive. In certain cases, countersunk bolts may also reduce the achievable resistance of the connection.

Structural bolts fall into two groups: 'ordinary' or 'black' bolts, and preloaded or high-strength friction grip (HSFG) bolts. All bolts are available in a range of lengths, available generally at 10 mm increments, and various diameters, but the industry standards for general application are 20 and 24 mm diameter grade 8.8 bolts. Fully threaded bolts are produced in a small range of lengths and, for M20 grade 8.8 bolts, the preferred length of bolt is 60 mm.

Preloaded or HSFG bolts are generally only used in applications where the bolt slip would otherwise impair the structural performance, or where the structure is exposed to dynamic or fatigue loading. HSFG bolts are produced from the same material as grade 8.8 bolts but have a larger head and stronger nut to avoid tensile failure during tightening by a torque wrench. Because a connection using HSFG bolts relies on friction between the steel interfaces rather than the bearing capacity of the bolt, a connection using HSFG bolts may require marginally more bolts than a similar connection using grade 8.8 bolts. The higher cost of HSFG bolts and their tightening procedure means that they are more expensive than grade 8.8 bolts.

Minimum and maximum bolt spacings and edge distances are specified in BS 5950 Part 1. These requirements for connections

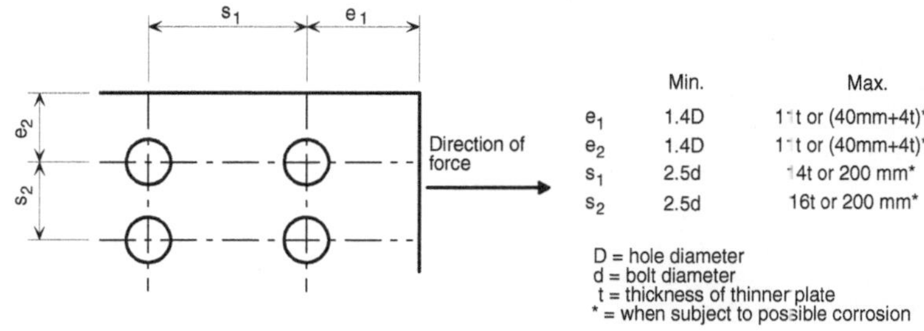

14.1 Minimum and maximum bolt spacings and edge distances in connections subject to shear forces

subject to shear in the bolts are summarised in Figure 14.1. Maximum edge distances are introduced only where there is a risk of corrosion between the plates. An example of a bolted connection with a shaped gusset plate is illustrated in Figure 6.22.

14.2 Welding

Most fabrication shops use manual metal arc (MMA), semi-automatic and fully automatic equipment, depending on the weld type and the length of run. The detailing of the steelwork must take account of the welding procedure. The welding processes that are commonly used are shown in Table 14.1.

Welding equipment has become much lighter, and more readily portable and easier to use in site applications. Normal MMA welding techniques are usually employed for site welding, although preheat

Table 14.1 Commonly used welding processes

Process	Automatic or manual	Shielding	Main use	Workshop or site	Comments	Maximum size fillet weld in single run
Manual metal arc (MMA)	Manual	Flux coating on electrode	Short runs	Workshop or site	Fillet welds larger than 6 mm are usually multi-pass, and are relatively expensive	6 mm
Submerged arc (SUBARC)	Automatic	Powder flux deposited over arc and recycled	Long runs or heavy butt welds	Workshop or occasionally on site	With twin heads, simultaneous welds either side of joint are possible	10 mm
Metal inert gas (MIG)	Automatic or semi-automatic	Gas (generally CO_2)	Short or long runs	Workshop	MIG has replaced manual welding in many workshops	8 mm

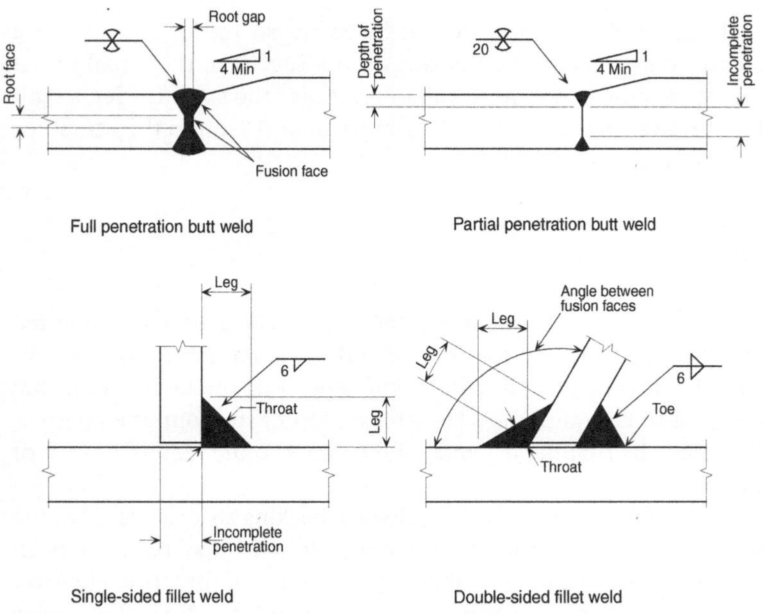

14.2 Butt and fillet welds between steel plates

may be required for some types and thicknesses of steel. Overhand or downhand welding is preferred because of the difficulty (and hence the slow speed) of underhand welding. Situations with underhand site-welding should be eliminated by careful design.

14.2.1 *Types of welds*

Two generic forms of welded connection between steel plates and steel sections are butt and fillet welds, which are illustrated in Figure 14.2.

Fillet welds are preferred in general construction because of their ease of placement and their lower heat input. Symmetrically placed fillet welds reduce angular distortion when welding plates. Single-sided fillet welds are usually only used in connections between hollow sections. Control of distortion is often left to the skill of the fabricator. Butt welds require preparation of the ends of the elements to be welded and are therefore more expensive. However, they can achieve full strength of the connected elements.

Welds always have a characteristic rippled surface. Sometimes it is requested that they should be ground down to a smooth finish for aesthetic or structural reasons. Welds suitable for grinding down are made a little larger, as metal is lost in grinding and the strength has to be maintained. Of course this involves additional work and cost. Specifications frequently state 'All welds to be ground down'. It may be possible to make worthwhile savings by specifying 'All visible welds to be ground down', unless there are structural reasons for removing surface defects, for example in a fatigue sensitive connection.

14.2.2 *Fillet welds*

Fillet welds are generally used to attach end-plates or brackets. Fillet welding is usually cheaper than butt welding but it relies on projections (of up to 15 mm) to be able to adequately place the weld.

A fillet weld can be specified by its throat thickness and/or leg length. Where the throat thickness is not specified, the actual throat thickness should not be less than 0.7 times the specified leg length for concave fillet welds, and not more than 0.9 times the actual leg length for convex fillet welds.

14.2.3 Butt welds

Butt welds join all or part of the cross-section of the connected members and may be specified as full or partial penetration welds. Partial penetration is often sufficient depending on the loads that are required to be transferred. The preparation of the connected parts is first made by shaping the ends of the plates to the required depth of the weld.

Where full penetration is required, a 'backing strip' is placed at the remote side of the weld. Alternatively, the weld may be made from both sides by preparing or shaping both sides of the plate. However this is usually more costly than single-sided welding. Butt welds should generally be made in the factory, as quality control is difficult to achieve on-site.

14.2.4 Welding conditions

Welding should be carried out in workshops or under controlled conditions. Precautions should be taken during site welding to protect the workpiece from adverse weather conditions. When surfaces to be welded are wet, or where the ambient temperature is O°C or less, the sections must be heated locally until they are warm to the touch for a distance of not less than 75 mm on either side of the joint. Furthermore, the surfaces to be welded should be dry, clean and free from rust, oil, grease and paint (except 'weld through' primer).

When subjected to elevated temperatures during welding or cutting, fumes will be produced which may be injurious to health. To ensure that threshold limits are not exceeded, the primer paint should be removed in the area which is to be treated, or provision should be made for adequate ventilation during welding or cutting which, if necessary, include local fume extraction.

14.3 Complex welded node of tubular column

14.3 Welding tubular sections

Structural hollow sections are made from steels complying with the requirements of BS EN 10 210[34] and are thus of weldable quality. In order for the steel to be weldable, it is important that, at the time of order, the steel is specified as having a maximum carbon equivalent value as specified in Table A2 or B2 of BS EN 10 210. A variety of weld details may be employed for tubular members and some assemblies can require complex welding, for example as with the connection between the tubular column and its inclined arms in Figure 14.3.

In some special applications, distortion of the cross-section due to the welding process should be carefully controlled, and it is necessary to seek the fabricator's advice as to the appropriate details and welding specification.

14.3.1 Manual metal arc welding

Manual metal arc (MMA) welding was the welding process most commonly used for SHS construction. Its use is being replaced by the semi-automatic or metal inert gas (MIG) welding process which is becoming the more popular, especially for shop-welded fabrication. MMA is often the common process employed in cases of restricted access.

14.3.2 Fillet and fillet-butt welds

Fillet-butt welds are used when welding CHS or RHS members, as the ends of the incoming section may be profiled. The terms describe the welding conditions which apply when various size ratios of bracing to the main member are involved.

Figure 14.4 shows the weld details for two bracings meeting main members of considerably different sizes. In both cases, welding

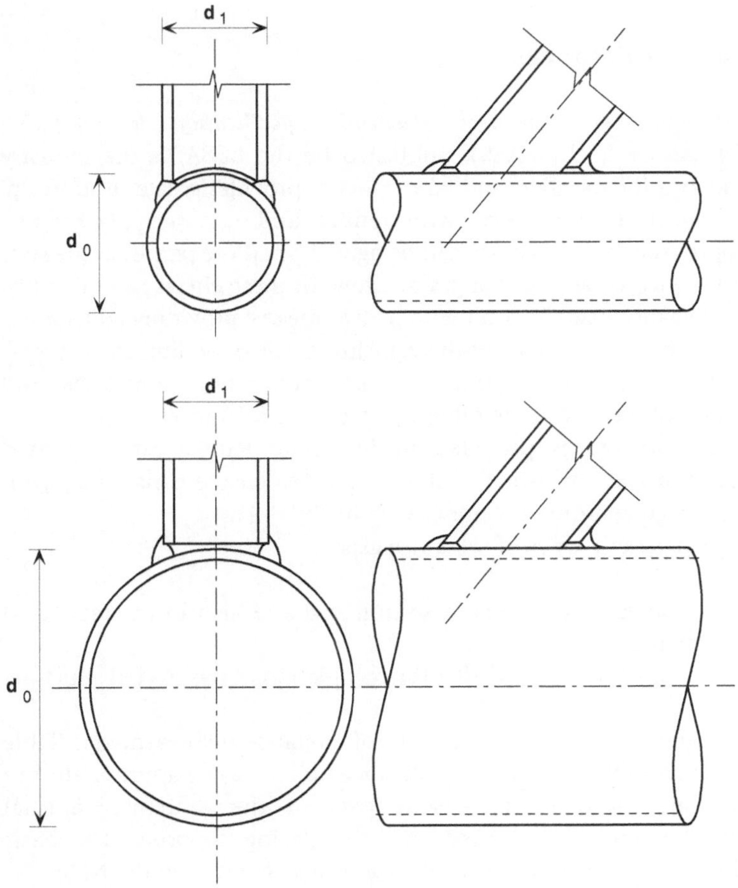

14.4 Fillet and fillet-butt welds to CHS members

conditions at the crown are similar and so, for the same loads, identical fillets would be used. However, conditions differ at the side of the tube. The curvature of the larger main member gives good conditions for fillet welds, whilst the curvature of the smaller main member necessitates a butt weld. The change from fillet to butt welding must be continuous and smooth.

For calculating weld sizes, both types are considered as fillet welds. The fillet-butt preparation is used where the diameter of the bracing is one-third or more of the diameter of the main member.

14.3.3 Continuous welding

Continuous welding is preferred when connecting SHS or RHS members, even if not required structurally. This is advised because of the possible risk of corrosion in the gaps between the elements.

The usual practice in the fabrication of frames using tubular sections is to work towards the open ends, that is to start welding in the middle of a frame and to work outwards on alternate sides to the ends. This tends to reduce distortion and avoid cumulative errors.

Flange connections are usually associated with a tight tolerance on the length of the component. It is good practice to first complete all the other welding before locating and welding on the flanges as a final operation.

14.4 Tolerances

The *National Structural Steelwork Specification for Building Construction*[56] (the NSSS), published by the BCSA, is the industry standard for specifiers and fabricators. It provides greater uniformity in specifications issued with tender and contract documents. Appropriate tolerances should be agreed by all the parties at the start of the project, as late changes to allow for these effects can be costly.

Deviations can be dealt with by two means: providing adjustment to overcome local lack-of-fit (e.g. shims), or providing clearance to give two adjacent components separate zones. Both adjustment and clearance provisions are often referred to as tolerances.

The tolerances provided in BS 5950: Part 2 are permitted deviations set on behalf of the designer to ensure consistency with the design assumptions underlying BS 5950: Part 1.

Two generic types of tolerance exist:

- geometric variations in section size and also in the fabricated members
- member deviations after the completion of the erected structure.

Limitations on these two forms of tolerance are presented in Table 14.2 and Table 14.3. Stricter tolerance limits may be appropriate, but this may lead to increased costs of the structure. Special considerations may be required for glazing supports. The basic dimensional tolerances for rolled sections are given in the NSSS.[56]

Table 14.2 Typical geometric tolerances for steel sections (taken from the *National Structural Steelwork Specification for Building Construction*[61])

Criterion for tolerance		Acceptable tolerances for section type	
		Rolled UB/UC	Fabricated section
Depth of section, D (in mm)			
	$D \leq 180$	+3.0 or −2.0 mm	±4.0 mm
	$180 < D \leq 400$	+4.0 or −2.0 mm	
	$400 < D \leq 700$	+5.0 or −2.0 mm	
	$D > 700$	+6.0 or −2.0 mm	
Width of section, B (in mm)			
	$B \leq 110$	+4.0 or −1.0 mm	±4.0 mm
	$110 < B \leq 210$	+4.0 or −2.0 mm	
	$210 < B \leq 325$	+4.0 or −4.0 mm	
	$B > 325$	+6.0 or −5.0 mm	
Off-centre of web			
	$102 < D < 305$	3.2 mm	±5.0 mm
	$D > 305$	4.8 mm	
Bow of web			Greater of $d/150$ or 3 mm
Horizontality of flange			
	$B \leq 110$	1.5 mm	Greater of $B/100$ or 3 mm
	$B > 700$	2% of $b \not> 6.5$ mm	
Verticality of web at support			Greater of $d/300$ or 3 mm
Squareness of cut end not prepared for bearing (plan or elevation)		$D/300$	
Squareness of cut end prepared for bearing (plan or elevation)		$D/1000$	
Overall length		±2 mm	±3 mm
Straightness along length of beam		Greater of $L/1000$ or 3 mm	Greater of $L/1000$ or 3 mm

Note:

L member length

d web depth

D section depth

B section width (all in mm)

Table 14.3 Acceptable tolerances in general steel construction (*National Structural Steelwork Specification for Building Construction*[61])

Level of adjacent beams	± 5 mm
Position of floor beams at columns	± 10 mm
Level at each end of the same beam	5 mm
Position of beam from wall	±25 mm
Level of foundations	+0 to −3 mm
Position of holding down bolts	20 mm
Maximum gap between bearing surface of column end (depth D)	$D/1000 + 1$ mm

14.5 Deflections

Control of deflections and long-term movement is important if other elements that are attached to the structure may be adversely affected, or if the use of the building is impaired, such as by water leakage. It is not appropriate here to discuss the wide range of deflection limits that are appropriate for different applications, but only to note certain well-defined cases that the designer should address, as follows:

- edge beams supporting cladding, such as glazed façades
- sway deflection of unbraced frames subject to wind action
- vertical movement of low pitch portal frames or roofs
- deflection of beams supporting compartment walls
- visual deflections or installation of raised floors in long-span beams
- horizontal movement of tall structures
- limit to floor vibrations in specialist buildings, such as hospitals
- use of isolating pads for deflection-sensitive equipment
- movement of supports to cranes and travelling machinery.

The designer should agree sensible deflection limits with specialist suppliers of the cladding and lifts, etc., and with the steel fabricator. The normal limit on deflections for beams subject to imposed load is beam span/360,[37] in order that deflections are not noticeable and that partitions are not subject to cracking.

Stricter deflection limits may be required in many of the above cases. A total deflection limit of 60 mm is generally used for long span internal beams, subject also to limits on floor vibration. A limit of span/500 is often specified for edge beams supporting brittle forms of cladding. It may be necessary to pre-camber long-span beams (span > 12 m) in order to off-set permanent deflections which may be visually unacceptable. The amount of pre-camber is normally set at half the anticipated total deflection, but not less than a practical minimum of 25 mm.

Chapter 15

Other design considerations

15.1 Pre-contract involvement of the fabricator

The purpose and benefits of the pre-contract involvement of the steel fabricator are:

- the particular skills and experience of the fabricator in similar projects can be identified
- the fabricator can comment on the practicality and cost-effectiveness of the proposed scheme
- practical details and an appropriate degree of standardisation can be established
- the structural engineer and architect can incorporate comments by the fabricator, and update their designs drawings as necessary
- more accurate budget estimates can be prepared and monitored as the design evolves.

15.2 Drawing examination and approval

In terms of drawing submission, the architect sketches the details that are required visually. The structural engineer then develops these details and performs the structural calculations and refines them where necessary, liaising and coordinating with the architect, as appropriate.

The fabricator then produces detailed drawings for fabrication of the steelwork, based on the structural drawings, which may be supplied on disk. The major fabricators have computer-aided design (CAD) systems which can be programmed into shop machinery for fabrication.

Where the detailing is critical to the success of the scheme, the architect should insist on examining the chosen details for compliance with the design concept.

15.3 Key decisions/checklists

In traditional design solutions, where the steel is not exposed to view, the design and detailing of the connections is usually carried out by the steel fabricator, who will use details which are the easiest and the most economic to fabricate. In situations where the steel will be visible, this route may not lead to details acceptable to the architect. Changing the connection details at a late stage to suit the architectural requirements will be a costly exercise for all parties, including the structural engineer and fabricator. Therefore, typical connections should be designed, and details established, before the steelwork is tendered.

Essentially, the following check-list defines the appropriate criteria for these decisions:

- the details adopted must be structurally acceptable
- the structure and its details should be economic to fabricate
- the contractor (fabricator) may prefer to detail his own connections, but the designer should be confident that his design approach can be realised (this emphasises the need for a fabricator to be involved early in the project)
- the details should be aesthetically pleasing and elegant
- the details should lead to an efficient method of erection — the size of sub-assemblies and the location of splices/lifting cleats are important aspects of the erection procedure.

The designer should also define on the drawings and in the specification, the additional information required by the fabricator, such as:

- the design loads and moments for the connections
- if the connection is to be concealed or exposed
- whether or not the details be established solely by the fabricator
- if the connection is to be bolted or welded (depending also on aesthetic considerations)
- whether the welds should be ground smooth
- whether the welds can be intermittent (or, alternatively, they must be continuous)
- the method of removing the sag from tie rods, such as by turnbuckles, or the threaded ends of the rods and forks
- specialist aspects, such as whether to use couplers rather than turnbuckles (if acceptable to the architect). Turnbuckles can have a left-hand and right-hand thread to tighten or tension it. Turnbuckles are usually necessary in bars longer than 6 m.

The level of involvement of the fabricator in the design and detailing of the connections should also be established. This information is also required at the tender stage in order that the fabricator can estimate costs accurately.

15.4 Fabricator's responsibilities during erection

The structural designer should provide an initial erection method statement as part of the Construction (Design and Management) Regulations 1994.[4] The erector of the framework must ensure stability during erection, and, therefore, lifting lugs and other elements to be utilised during erection should be determined at the fabricator's shop. If the lifting lugs are designed to be permanent fixtures, the architect should liaise with the fabricator to ensure that they are acceptable aesthetically in the final design. Where lifting lugs are to be removed, provision should also be made for the removal, grinding down and reinstatement of protective coating.

15.5 Mock-ups and prototypes

Where complex or innovative details are proposed, it is often cost- and time-effective to construct a prototype or mock-up of the proposed construction. For example, a critical factor in the Waterloo International project was the nine months lead-in time before starting on site. This enabled the fabricator to make a full scale mock-up to test the assembly for buildability. As a result of this mock-up, the connection details were improved and some of the temporary work arrangements were revised. This helped the cladding and glazing sub-contractors, and highlighted sensible tolerances on installation. In particular, planar glazing requires very tight tolerances, which may require erection trials of the support structure.

The same principle was applied to the Law Faculty at Cambridge (see Figure 4.16). The project was originally designed with cast nodes, as it was envisaged that all castings would be the same. In the final design, this was not the case. The contractor offered a welded joint which was accepted by the designers. The glazing cleats on the nodes were produced by investment casting (or the lost wax process) to achieve a finer finish.

15.6 Transportation of steelwork

Current legislation requires that loads up to 5.1 m wide require one month's notice for transportation, subject to a maximum length of 27 m and a maximum height requirement of 4.93 m. Smaller loads up to 3.6 m wide require no specific notice (although they must observe the same height and length limits). Larger assemblies should be fabricated as sub-assemblies of less than these dimensions, which are connected together on site. As noted earlier, the connection and detailing of the interfaces between these sub-assemblies is very important to the aesthetics and adequacy of the final design.

Chapter 16

References and sources of information

This section reviews sources of information and useful contacts. It is not exhaustive, but is intended to be a useful first step.

Further sources of information

Castings
Casting Technology International

Composite construction
The Steel Construction Institute
Corus Construction Centre

Connections
The Steel Construction Institute
British Constructional Steelwork Association Ltd

Corrosion
W & J Leigh Ltd
International Paints
Corus (Swinden Technology Centre)

Curved sections
The Angle Ring Co. Ltd
Barnshaw Section Benders Ltd

Environmental issues
The Steel Construction Institute
Corus (Swinden Technology Centre)

Fire engineering
The Steel Construction Institute
Warrington Fire Research Centre Ltd
Corus (Swinden Technology Centre)
Building Research Establishment

Fire protection
Association of Structural Fire Protection Contractors (ASFPC)

Galvanizing
The Galvanizers' Association

Glazing
Pilkington Glass Ltd

Intumescent Coatings
W & J Leigh Ltd
International Paints
Nullifire

Quality assurance
British Constructional Steelwork Association Ltd

Slimdek
Corus Construction Centre
The Steel Construction Institute

Steel specification
The British Constructional Steelwork Association
Corus Construction Centre

Strip steel products
Corus Colors

Tubular construction
Corus Tubes and Pipes

Welding technology
The Welding Institute

Sources of advice

BRITISH CONSTRUCTIONAL STEELWORK ASSOCIATION LTD
4 Whitehall Court
Westminster
London
SW1A 2ES
Tel. 0207 839 8566
Fax. 0207 976 1634

CORUS CONSTRUCTION CENTRE
PO Box 1
Frodingham House
Scunthorpe
DN16 1BP
Advisory tel. line:
01742 405060

CORUS COLORS
Innovation Centre
Shotton Works
Shotton
Deeside
CH5 2NH
Tel. 01244 892 434

CORUS TUBES AND PIPES
Technical Marketing
PO Box 101
Corby
Northants
NN17 5DA
Free phone 0500 123 133
Tel. 01536 402121
Fax. 01536 404049

CORUS STRIP PRODUCTS
PO Box 10
Newport
Gwent
NP19 0XN
Tel. 01633 755 171

CORUS TECHNOLOGY
Swinden Laboratories
Moorgate
Rotherham
S. Yorks
S60 3AR
Tel. 01709 820166
Fax. 01709 825337

GALVANIZERS' ASSOCIATION
Wrens Court
56 Victoria Road
Sutton Coldfield
West Midlands
B72 1SY
Tel. 0121 355 8838
Fax. 0121 355 8727

THE STEEL CONSTRUCTION INSTITUTE
Silwood Park
Ascot
Berkshire
SL5 7QN
Tel. 01344 623345
Fax. 01344 622944

THE WELDING INSTITUTE
Abington Hall
Abington
Cambridge
CB1 6AL
Tel. 01223 891 162

Specialist companies

THE ANGLE RING CO. LTD
Bloomfield Road
Tipton
West Midlands
DY4 9EH
Tel. 0161 834 8441
Fax. 0161 832 4280

BARNSHAW SECTION BENDERS LTD
Tipton Road
Tividale, Warley
West Midlands
B69 3HY
Tel. 0121 557 8261
Fax. 0121 557 5323

CASTING TECHNOLOGY INTERNATIONAL
No. 7 East Bank Road
Sheffield
S2 3PT
Tel. 0114 272 8647
Fax. 0114 273 0852

PILKINGTON GLASS LIMITED
Pilkington Architectural
Alexandra Works
St Helens
WA10 3TT
Tel. 01744 692998
Fax. 01744 451326

Bibliography

The following documents contain useful information but are not referred to directly in the text

British Standards

BS EN 499: 1995
Welding consumables — general electrodes for manual metal arc welding of non alloy and fine grain steels — classification (Replaces BS 639: 1986)

BS 3692: 2001
ISO metric precision hexagon bolts, screws and nuts: metric units

BS 4190: 2001
ISO metric black hexagon bolts, screws and nuts

BS 4395: Part 1: 1969
High strength friction grip bolts and associated nuts and washers for structural engineering: metric series: general grade

BS 5531: 1988
Code of practice for safety in erecting structural frames

BS EN 10027- 1: 1992
Designation systems for steel: steel names, principal symbols

BS 6399:
Loading for Buildings, Parts 1 to 3

BS EN 10034: 1993
Structural steel I and H sections. Tolerances on shape and dimensions

ENV 1090:
Erection of steel structures

BS 3100:
Specification of steel castings for general engineering purpose

BS 476: Parts 20 and 21
Fire tests on building materials and structures

BS 8202:
Coatings for fire protection of building elements

BS 4479: 1990
Design of articles to be coated

BS EN 22553: 1995
Welded, brazed and soldered joints — symbolic representation on drawings (replaces BS 499: Part 20)

Other useful publications

Design in Steel Series
Various publications covering Adaptability, *Slimflor*, *Slimdek* and Cellular Beams.
Corus publications (available from the Corus Construction Centre)

Architecture and Construction in Steel
Editors: A. Blanc, M. McEvoy and R. Plank
Published 1993 E and FN Spon

The Art of the Structural Engineer
B. Addis. Published by Artemis 1994

Architectural Teaching Resource: Studio Guide, 2nd Edition
Published by The Steel Construction Institute, 2000

Architecture in Steel — The Australian Context
Alan Ogg
Published by The Royal Australian Institute of Architects, 1987

Structural Steelwork Design. Structural Design Guide to BS 5950: Part 1: 2000
Published by the SCI and BCSA, 2001, 6th Edition

Structural Steelwork — Fabrication
Volume One: Publication No. 7/80
B. J. Davies and E. J. Crawley
Published by the BCSA

Structural Steelwork — Erection
Publication No. 20/89
W. H. Arch
Published by the BCSA

Steel Construction Yearbook
MacMillan Scott, 2003

Welding Process Technology
P. T. Houldcroft
Cambridge University Press, Cambridge, 1977

Steel Designer's Manual
6th Edition
Blackwell Scientific, 2003

Structural Detailing in Steel
M.Y.H. Bragasl
Thomas Telford, 2000

Steel Detailers' Manual
A. Hayward and F. Weare
Blackwell Science, 2002

RIBA CPD Packages in the Open Learning Series:
Tension Structures by John Thornton. See AJ 16 Sept 1992
The Integration of Structure and Services in Long Span Commercial Buildings by Peter Trebilcock. See AJ 30 Sept 1992
Corrosion Protection of Steelwork by Ken Johnson. See AJ 30 Sept 1992
The Fabrication and Erection of Steelwork by Peter Trebilcock

Brick Cladding to Steel Framed Buildings
Published jointly by the Brick Development Association and British Steel, 1986

Structural Steelwork Fabrication
R. Taggart. Corus Construction Centre (see also reprint from *The Structural Engineer*, Vol. 64A, No. 8. August 1986)

Light Steel and Modular Construction
Building Design Using Cold Formed Steel: Construction, Detailing and Practice
SCI Publication P165
Modular Construction Using Light Steel Framing: An Architect's Guide
SCI Publication P272
Corrosion Protection Guide for Steelwork in Building Refurbishment
Corus publications

Fire Protection for Structural Steel in buildings
3rd Edition
Published by the ASFPC and the SCI

Structural Fire Engineering
Investigation of Broadgate Phase 8 Fire
SCI Publication

Structural Fire Safety: A Handbook for Architects and Engineers
SCI Publication P197, 1999

Architect's working details

Architect's Working Details 1
Edited by David Jenkins and Louis Dezart

Architect's Working Details 2
Edited by Alastair Blyton

Architect's Working Details 3
Edited by Susan Dawson

Architect's Working Details 4
Edited by Susan Dawson

Architect's Working Details 5
Edited by Susan Dawson

Architect's Working Details 6
Edited by Susan Dawson

Architect's Working Details 7
Edited by Susan Dawson

All published by EMAP Construct.

Relevant case studies dealing with steel projects or useful details from these seven volumes are presented in the following list (see overleaf).

Page	Title	Architect
Architect's Working Details 4		
13	Adaptation of a proprietary glazing system, BRIT, Performing Arts and Technology School, Croydon	Cassidy Taggart Partnership
45	Double doors, Sculpture Gallery, Leeds	Jeremy Dixon, Edward Jones
59	Glazing & support structure to a vaulted atrium, Shopping Centre, Norwich	Lambert Scott & Innes
75	A glazed steel roof & wall structure, Station, East Croydon	Alan Brookes Associates
79	A roof of air filled cushions Schlumberger Building, Cambridge	Michael Hopkins & Partners
87	A curved profiled sheet, metal roof, Surgery, Swiss Cottage, London	Pentarch Limited
91	Space frame Vault, Bentall Centre, Kingston-upon-Thames	Building Design Partnership
99	A timber & steel church roof, Brentford, Middlesex	Michael Blee Design
103	Glazed wall and roof light, Factory, Glasgow	The Ryder Nicklin Partnership
111	A barrel-vaulted plant enclosure, Public Library, Croydon	Tibbalds Monro
118	A fabric roof to a stairwell, Regent St, London	Sheppard Robson
121	An asymmetric fabric roof, Channel Tunnel Terminal, Sussex	Building Design Partnership
125	Pitched roof with extended eaves, University, Enfield, Middlesex	Rivington Street Studio
135	Balcony front & lighting auditorium, Southend-on-Sea, Essex	Tim Foster Architects
143	Steel staircase and balustrade, Horndean Community School, Hampshire	Hampshire County Council Architects' Department
151	A staircase with glazed balustrade, Theatre, Edinburgh	The Law and Dunbar-Nasmith Partnership
166	A freestanding glazed umbrella canopy, Sydney, Australia	Harry Seidler and Associates
169	A glazed footbridge, Car Park, Sunderland	Napper Collerton Partnership
Architect's Working Details 5		
10	Church Hall extension, Pinner	Weston Williamson
22	Acoustically isolated external wall	Simpson Associates
24	Triangular panels on a curved facade	Alsop & Stormer
32	External steelwork to pavilion	Studio Downie
36	Glass rainwater conductor	Nicholl Russell studios
44	Glazed wall and supporting sun-screen	Studio BAAD
52	External steelwork in St Catherine's Oxford	Hodder Associates
64	Glazed facade and curved roof truss, Cardiff Bay	Ahrends, Burton & Kovalck
72	Curved glass wall and tubular Vierendeel truss support	Foster and Partners
87	Steel framed wind tower, Ionica, Cambridge	RH Partnership
116	Cast structural nodes, Queen's Dock, Liverpool	David Marks, Julia Barfield
132	Steel bridge with slate steps, Cable & Wireless, Coventry	MacCormac Jamieson Pritchard
144	Steel stairs, Channel 4 building	Richard Rogers Partnership
156	Glazed canopy, UEA, Norwich	John Miller & Partners
160	Student centre, University of Liverpool	King McAllister
172	Stainless steel lobby	Hallet & Pollard
Architect's Working Details 6		
7	Wall and gallery of a recital room	Van Heynigen & Howard
23	Detailing profiled metal cladding	Paul Hyatt Architects
71	An undulating glazed roof	Wintersgill & Faulkner
79	A curved glass roof/canopy	DEGW
83	A spectacular day lit structure founded on old warehousing	Munkenbeck & Marshall
91	Junctions between concrete and steel roof structures	Richard Rogers Partnership
95	An inflated roof structure	David Morley Architects
107	An external steel safety staircase	CZWG

Page	Title	Architect
Architect's Working Details 6 (continued)		
131	Inserting a mezzanine floor	Associated Architects
151	Canopy for petrol station	Lifshutz Davidson
155	Artists of the floating bridge	Future Systems
159	An outside room with a view	Anthony Grimshaw Associates
Architect's Working Details 7		
19	A glass-fibre clad wall and glazed clerestory	Chris Wilkinson Architects
55	A touch of class	CZWG
79	A curved metal roof with an eaves clerestory	Aldington Craig & Collinge
87	A steel staircase with a cast-glass balustrade	McDowell and Benedetti
99	A glass and steel staircase	Levitt Burnstein Associates
127	Radical intervention	Studio Downie
131	A swimming pool structure	Studio E Architects
135	A steel structure with 'tree' columns	Bennetts Associates
155	Taken from a riverbank	Brookes Stacey Randall

Corus publications on tubes

TD references refer to Corus publications available free of charge from Corus, Tubes & Pipes, Corby:

TD 167
Hot finished RHS and CHS Sizes, Properties and Technical Data: BS 4848
Lists the size range available in hot finished and gives the technical information, geometric properties and tolerances.

TD 328
SHS Welding: BS 5135
Describes electric arc welding techniques applicable to RHS and CHS.

TD 325
SHS Jointing
Describes numerous methods of jointing structural hollow sections.

TD 349
Design in SHS: BS 5950
A brochure for engineers to assist them in using SHS limit state design to BS 5950: Part 1: 1990.

TD 338
Design of SHS Welded Joints: BS 5950
A manual to simplify the design of predominantly axially loaded welded tubular joints compatible with the requirements of BS 5950: Part 1: 1990.

TD 359
Design of Lattice Roof Structures in Structural Hollow Sections
A basic introduction into lattice roof structures for framed industrial buildings using SHS.

TD 306
An Introduction to Corus Tubes Electric Welded Hot Formed Structural Hollow Sections
Describes the manufacture, quality control, specifications, tolerances, size range and properties of welded SHS and lists comparable international standards.

TD 304
Quality Large Diameter SAW Pipe
Describes the submerged arc welded manufacturing process and the specifications available for large diameter pipes beyond the limit of the BS 4848 range, and illustrates their use in typical conveyance, piling and structural applications.

TD 296
Design Manual for SHS Concrete Filled Columns — Part 1 and Part 2
Design manual for calculating the load bearing capacity for both CHS and RHS columns (Part 1).
Design manual for calculating the fire resistance of RHS columns (Part 2).

TD 361
Design for SHS Fire Resistance to BS 5950: Part 8
Presents the different methods that can be used to obtain fire resistance with hot finished SHS.

TD341
Cold form
Describes the manufacture and quality control of cold-formed SHSs and RHSs to BS 6363, and gives the size, range and technical information, including geometric properties to BS 6363 and axial capacities.

CIDECT Publications on Construction with Hollow Steel Sections (all published by Verlag TUV Rheinland, but available from the SCI).

1. *Design Guide for Circular Hollow Section (CHS) Joints under Predominantly Static Loading.*
2. *Structural Stability of Hollow Sections.*
3. *Design Guide for Rectangular Hollow Sections (RHS) under Predominantly Static Loading.*
4. *Design Guide for Structural Hollow Section Columns Exposed to Fire.*
5. *Design Guide for Concrete Filled Hollow Section Columns under Static and Seismic Loading.*

Other CIDECT publications:

Tubular Structures in Architecture. M. Eekhout, University of Delft (available from the SCI).

References

Note:
SCI refers to the Steel Construction Institute.

1
Interfaces: Curtain Wall Connections to Steel Frames
SCI Publication P101

2
Interfaces: Connections Between Steel and Other Materials
SCI Publication P102

3
Interfaces: Electric Lift Installations in Steel for Buildings
SCI Publication P103
SCI/NALM

4
Brown D.G.
The Construction (Design and Management) Regulations 1994: Advice for Designers in Steel
SCI Publication P162

5
Joints in Simple Construction
The Steel Construction Institute/British Constructional Steelwork Association
SCI Publication P212

6
Joints in Steel Construction Moment Connections
The Steel Construction Institute/British Constructional Steelwork Association
SCI Publication P207

7
Trebilcock P.J. and Lawson R.M.
Buildings Design Using Cold Formed Steel Sections: An Architect's Guide
SCI Publication 130

8
Lawson R.M., Grubb P.J., Prewer J. and Trebilcock P.
Modular Construction Using Light Steel Framing: An Architect's Guide
SCI Publication P272

9
King C.M. and Brown D.G.
Design of Curved Steel
SCI Publication P281

10
Lawson R.M.
Design of Composite Slab and Beams with Steel Decking
SCI Publication P55

11
McKenna P.D. and Lawson R.M.
Design of Steel Framed Buildings for Service Integration
SCI Publication P166

12
Ward J.K.
Composite and Non-Composite Design of Cellular beams
SCI Publication P100

13
Mullett D.L.
Design of Slimflor Construction
SCI Publication P110

14
Mullett D.L. and Lawson R.M.
Design of Slimflor Fabricated Beams using Deep Composite Decking
SCI Publication P248

15
Lawson R.M, Mullet D.L. and Rackham J.W.
Design of Asymmetric Slimflor Beams with Deep Composite Decking
SCI Publication P175

16
The Slimdek Manual
Corus Construction Centre, 2000

17
Mullett D.L.
Design of RHS Slimflor Edge Beams
SCI Publication P169

18
Lawson R.M.
Design of openings in the webs of Composite Beams
SCI/CIRIA Publication P68

19
Lawson R.M. and McConnel R.E.
Design of Stub Girders
SCI Publication P118

20
Owens G.W.
Design of Fabricated Composite Beams in Buildings
SCI Publication P59

21
Lawson R.M and Rackham J.W
Design of Haunched Composite Beams in Buildings
SCI Publication P60

22
Neal S. and Johnson R.
Design of Composite Trusses
SCI Publication P83

23
Brett P. and Rushton J.
Parallel Beam Approach — A Design Guide
SCI Publication P74

24
Newman G.M.
The Fire Resistance of Web Infill Steel Columns
SCI Publication P124

25
Chung K.F. and Narayanan R.
Composite Column Design to Eurocode 4
SCI Publication P142

26
Flowdrill
Corus Tubes and Pipes

27
Architecture and Construction in Steel
Chapter 20: 'Tensile Structures'
Cheynan and Hill Publishers

28
Makowski, Z.S.
Analysis Design and Construction of Double Layer Grids
Applied Science Publisher, 1981

29
Steel Supported Glazing Systems
SCI Publication P193

30
Structural Use of Glass
Institution of Structural Engineers, 1999

31
Baddoo N.R., Burgan B. and Ogden R.G
Architects' Guide to Stainless Steel
SCI Publication P179

32
BS EN 10113: 1993
Hot-rolled products in weldable fine grain structural steels

33
BS EN 10025: 1993
Hot-rolled products in non-alloy structural steels — technical delivery conditions

34
BS EN 10210-1: 1994
Hot finished structural hollow sections of non-alloy and fire grain structural steels

35
BS EN 10219: 1997
Cold formed weldable structural steel sections of non-alloy and fine grain steels

36
BS EN 10155:
Structural steels with improved atmospheric corrosion resistance

37
BS 5950: Part 1: 2000
Structural use of steelwork in building: Code of Practice for design in simple and continuous construction: Hot rolled sections

38
prEN1993-1-1: Eurocode 3
Design of steel structures Part 1.1: 1993
General rules and rules for building

39
BS 5950: Part 3: 1990
Structural use of steelwork in building: Code of Practice for design in composite construction: Section 3.1 Composite beams

40
prEN1994-1-1: Eurocode 4
Design of composite steel and concrete structures Part 1.1: General rules and rules for building

41
BS 5950: Part 8: 2000
Structural use of steelwork in building: Code of Practice for fire resistant design

42
prEN1993-1-2: Eurocode 3
Design of steel structures Part 1.2: Structural fire design

43
prEN1994-1-2: Eurocode 4
Design of composite steel and concrete structures Part 1.2: Structural fire design

44
BS EN 10088:
Stainless steels

45
prEN1993-1-4: Eurocode 3
Design of steel structures: Part 1.4: General rules; Supplementary rules for stainless steels

46
Baddoo N.R
Castings in Construction
SCI Publication P172

47
The Prevention of Corrosion on Structural Steelwork
Corus (former British Steel) Publication

48
BS EN ISO 14713: 1999
Protection against corrosion of iron and steel in structures. Zinc and aluminium coatings. Guidelines

49
BS EN ISO 12944: 1998
Paints and varnishes. Corrosion protection of steel structures by protective paint systems

50
BS 7079: 1990
Preparation of steel substrates before application of paints, (also ISO 8501-1)

51
BS EN 8501: 2001
Preparation of steel sub-strates before application of paints and related products

52
BS EN 8504-2: 2001
Preparation of steel sub-strates before application of paints and related products; Abrasive blast cleaning

53
BS EN ISO 1461: 1999
Hot dip galvanized coatings on fabricated iron and steel articles. Specifications and test methods

54
BS EN 22063: 1993
Metallic and other inorganic coatings. Thermal spraying. Zinc, aluminium and their alloys

55
Fire Protection for Structural Steel in Buildings
Association of Specialist Fire Protection Contractors and Manufacturers
3rd Edition

56
BS 8110 Part 1
The Structural Use of Concrete, 1997

57
Yandzio E., Dowling J.J and Newman G.M.
Structural Fire Design: Off-Site Applied Thin Film Intumescent Coatings
SCI Publication P160

58
Bond G.V.L.
Fire and Steel Construction: Water Cooled Hollow Columns
SCI Publication P38

59
Newman G.M., Robinson J.T and Bailey C.G.
Fire Safe Design: A New Approach to Multi-storey Framed Buildings
SCI Publication P288

60
Law M.R. and O'Brien T.P.
Fire Safety of Bare External Structural Steel
SCI Publication P09

61
National Structural Steelwork Specification for Building Construction (3rd Edition)
British Construction Steelwork Association.

62
BS 5950: Part 2
Structural use of steelwork in buildings: Specification for materials, fabrication and erection: Hot-rolled sections

Index